AF358678

Signals and Images in Sea Technologies

Signals and Images in Sea Technologies

Editors

Davide Moroni
Ovidio Salvetti

MDPI • Basel • Beijing • Wuhan • Barcelona • Belgrade • Manchester • Tokyo • Cluj • Tianjin

Editors

Davide Moroni	Ovidio Salvetti
Institute of Information Science and Technologies	Institute of Information Science and Technologies
National Research Council of Italy	National Research Council of Italy
Pisa	Pisa
Italy	Italy

Editorial Office
MDPI
St. Alban-Anlage 66
4052 Basel, Switzerland

This is a reprint of articles from the Special Issue published online in the open access journal *Journal of Marine Science and Engineering* (ISSN 2077-1312) (available at: www.mdpi.com/journal/jmse/special_issues/signals_images).

For citation purposes, cite each article independently as indicated on the article page online and as indicated below:

LastName, A.A.; LastName, B.B.; LastName, C.C. Article Title. *Journal Name* **Year**, *Volume Number*, Page Range.

ISBN 978-3-0365-1356-0 (Hbk)
ISBN 978-3-0365-1355-3 (PDF)

Contents

About the Editors . . . . **vii**

Davide Moroni and Ovidio Salvetti
Signals and Images in Sea Technologies
Reprinted from: *Journal of Marine Science and Engineering* **2021**, *9*, 41, doi:10.3390/jmse9010041 . . **1**

Kai Hu, Yanwen Zhang, Feiyu Lu, Zhiliang Deng and Yunping Liu
An Underwater Image Enhancement Algorithm Based on MSR Parameter Optimization
Reprinted from: *Journal of Marine Science and Engineering* **2020**, *8*, 741, doi:10.3390/jmse8100741 . **5**

Diana Vargas, Ravindra Jayaratne, Edgar Mendoza and Rodolfo Silva
On the Estimation of the Surface Elevation of Regular and Irregular Waves Using the Velocity
Field of Bubbles
Reprinted from: *Journal of Marine Science and Engineering* **2020**, *8*, 88, doi:10.3390/jmse8020088 . . **29**

Denis Selimović, Jonatan Lerga, Jasna Prpić-Oršić and Sasa Kenji
Improving the Performance of Dynamic Ship Positioning Systems: A Review of Filtering and
Estimation Techniques
Reprinted from: *Journal of Marine Science and Engineering* **2020**, *8*, 234, doi:10.3390/jmse8040234 . **45**

Mario D'Acunto, Davide Moroni, Alessandro Puntoni and Ovidio Salvetti
Optimized Dislocation of Mobile Sensor Networks on Large Marine Environments Using
Voronoi Partitions
Reprinted from: *Journal of Marine Science and Engineering* **2020**, *8*, 132, doi:10.3390/jmse8020132 . **73**

Mengchen Xiong, Jie Zhuo, Yangze Dong and Xin Jing
A Layout Strategy for Distributed Barrage Jamming against Underwater Acoustic Sensor
Networks
Reprinted from: *Journal of Marine Science and Engineering* **2020**, *8*, 252, doi:10.3390/jmse8040252 . **85**

Yu Hao, Nan Zou and Guolong Liang
Robust Capon Beamforming against Steering Vector Error Dominated by Large
Direction-of-Arrival Mismatch for Passive Sonar
Reprinted from: *Journal of Marine Science and Engineering* **2019**, *7*, 80, doi:10.3390/jmse7030080 . . **103**

About the Editors

Davide Moroni

Davide Moroni received an M.Sc. degree (Hons.) in Mathematics from the University of Pisa, in 2001, a Diploma from the Scuola Normale Superiore of Pisa, in 2002, and a Ph.D. degree in Mathematics from the University of Rome La Sapienza, in 2006. He is a Researcher with the Institute of Information Science and Technologies (ISTI), National Research Council, Italy, Pisa. He is currently the Head of the Signals and Images Lab, ISTI. He is the Chair of the MUSCLE working group (https://wiki.ercim.eu/wg/MUSCLE) of the European Consortium for Informatics and Mathematics. Since 2018, he has served as the Chair of the Technical Committee 16 on Algebraic and Discrete Mathematical Techniques in Pattern Recognition and Image Analysis of the International Association for Pattern Recognition (IAPR). He is an Associate Editor of IET Image Processing. His main research interests include geometric modeling, computational topology, image processing, computer vision, and medical imaging.

Ovidio Salvetti

Ovidio Salvetti (Ph.D.), Research Associate at ISTI-CNR, works in the fields of image analysis and understanding, multimedia information systems, spatial modeling, and intelligent processes in computer vision and information technology. He is the co-author of seven books and monographs and more than four hundred technical and scientific articles. He is the owner of eleven patents regarding systems and software tools for real-time signal and image acquisition and processing. He has been the scientific coordinator of several national and European research and industrial projects, in collaboration with Italian and foreign research groups, in the fields of computer vision, multimedia semantics, and high-performance computing for diagnostic imaging.

Journal of
Marine Science and Engineering

Editorial

Signals and Images in Sea Technologies

Davide Moroni and **Ovidio Salvetti ***

Istituto di Scienza e Tecnologie dell'Informazione "Alessandro Faedo" CNR, 56124 Pisa, Italy;
davide.moroni@isti.cnr.it
* Correspondence: ovidio.salvetti@isti.cnr.it

Citation: Moroni, D.; Salvetti, O. Signals and Images in Sea Technologies. *J. Mar. Sci. Eng.* **2021**, *9*, 41. https://doi.org/10.3390/jmse9010041

Received: 1 December 2020
Accepted: 24 December 2020
Published: 4 January 2021

Publisher's Note: MDPI stays neutral with regard to jurisdictional claims in published maps and institutional affiliations.

Life below water is the 14th Sustainable Development Goal (SDG) envisaged by the United Nations and is aimed at conserving and sustainably using the oceans, seas and marine resources for sustainable development [1]. It is not difficult to argue that *Signals and Image technologies* may play an essential role in achieving the foreseen targets linked to SDG 14. Indeed, besides increasing general knowledge of ocean health by means of data analysis, methodologies based on signal and image processing can be helpful in environmental monitoring [2], in protecting and restoring ecosystems [3], in finding new sensor technologies for green routing and eco-friendly ships, in providing tools for implementing best practices for sustainable fishing, as well as in defining frameworks and intelligent systems for enforcing sea law and making the sea a safer and more secure place. Imaging is also a key element for the exploration of the underwater world for various scopes, ranging from the predictive maintenance of sub-sea pipelines and other infrastructures to the discovery, documentation and protection of the sunken cultural heritage.

The main scope of this Special Issue has been to investigate the techniques and ICT approaches, and in particular the study and application of signal- and image-based methods and, in turn, to explore the advantages of their application to the main areas mentioned above. After a careful review process, six papers were included in the Special Issue. Their thematic contributions are discussed in below.

As is well known, vision in the underwater environment is a much more complicated task than it is overland. Methods that achieve good performance in an outdoor scene may completely fail when applied to underwater optical images. The sun radiation only penetrates a few meters into the water medium, and it undergoes an attenuation process that reduces the radiation components in an irregular and non-homogeneous way, depending on the light wavelength, resulting in a severe distortion of the spectral content of the image. Therefore, in many cases, image enhancement is an essential step in several image processing and computer vision pipelines addressing the undersea world. This is precisely the scope of the paper "An Underwater Image Enhancement Algorithm Based on MSR Parameter Optimization" by Hu at al. [4]. Among the general-purpose image enhancement approaches, algorithms based on the so-called Multi-Scale Retinex (MSR) have achieved good results for overland images. However, they cannot be applied directly to the underwater domain since the general parameters cannot be adapted to the submerged scenario. The authors of the paper propose an underwater image enhancement optimisation (MSR-PO) algorithm, which uses the non-reference image quality assessment (NR-IQA) index as the optimisation index. To this end, they perform several experiments that make it possible to determine the appropriate parameters in MSR as the optimisation object. The reported results show that the proposed algorithm has unparalleled adaptability to disparate submarine scenes.

Image analysis plays an important role not only at sea but also for the characterisation of vessels and floating vehicles in water flume, an important topic in naval engineering. In this context, imaging can be used to determine the velocity field of bubbles, providing a non-intrusive technique for the kinematic study of the Free Surface Elevation of Water (FSEW). In the paper by Vargas et al. [5] entitled "On the Estimation of the Surface Elevation of Regular and Irregular Waves Using the Velocity Field of Bubbles", imaging is used to

1

infer the wave velocities and forces acting on maritime infrastructure, one of the most challenging topics in the field of Coastal Engineering and Fluid Dynamics. In the study, a curtain of artificial bubbles was generated by a compressed air device placed at the bottom of the flume. Since the velocity of the bubbles can be considered as a proxy of the water velocity, particle imaging velocimetry (PIV) allows for an estimation of the velocity of the bubbles and in turn of the FSEW by using a transfer function.

Waves also intervene in ships dynamics and positioning. Indeed, the natural forces of induced nonlinear waves acting on a ship's hull interfere with the ship movements, possibly leading to sub-optimal routes, wrong positioning and safety concerns. There has been a growing interest in Dynamic positioning (DP) systems that consist of a complex network of sensors and actuators for obtaining accurate positioning. However, the sensor data is usually noisy, with scarce accuracy and reliability for reconstructing directional wave spectra; therefore, suitable filtering and signal processing appear to be necessary. The paper "Improving the Performance of Dynamic Ship Positioning Systems: A Review of Filtering and Estimation Techniques" by Selimović et al. [6] aims at presenting the reader with an overview of the state-of-the-art estimation and filtering techniques for providing optimum estimation states in DP systems, with a view towards safe and green routing.

Aspects related to the protection of the environment are also the main driver of the paper "Optimised Dislocation of Mobile Sensor Networks on Large Marine Environments Using Voronoi Partitions" by D'Acunto et al. [7]. In this paper, solutions are sought for the dynamic dislocations of sensors, mainly e-noses and other AUV-borne devices, in order to promptly detect and respond to pollution events, thus reducing the costs for real-time surveillance of large areas. Indeed, generally, the probability of the occurrence of a pollution event depends on the burden of possible sources operating in the areas to be monitored. The paper presents theoretical and simulated results inherent to a Voronoi partition approach for the optimised dislocation of a set of mobile wireless sensors with circular (radial) sensing power on large areas. The optimal deployment was found to be a variation of the generalised centroidal Voronoi configuration.

Protection against adversarial Underwater Acoustic Sensor Networks (UASN) is the central theme of the paper by Xiong et al. entitled "A Layout Strategy for Distributed Barrage Jamming against Underwater Acoustic Sensor Networks" [8]. Distributed barrage technology can be used to prevent a target from being detected. The paper focuses on the capacity of a limited set of jammers to achieve the best jamming effect by maximising the Cramér–Rao bound of multiple targets estimates. A heuristic algorithm is used to solve this optimisation model, which is demonstrated thanks to several numerical simulations. Stability of the results with respect to errors in the positioning of the jammers is also discussed since it is of vital importance for practical applications.

Acoustic signal processing is also the main subject of the last paper entitled "Robust Capon Beamforming against Steering Vector Error Dominated by Large Direction-of-Arrival Mismatch for Passive Sonar" by Hao et al. [9]. Arrays of sensors, that is, sets of sensors used in a particular geometric pattern, are used primarily when it is necessary to increase the performance of the sensor or add new dimensions to the observation estimating more parameters, for instance, the Direction-of-Arrival (DoA) of impinging waves. In this context, adaptive beamforming is often used in passive sonar arrays to improve the detectability of weak sources. Historically, the main push to this research topic has been military applications, in particular, the acoustic detection of submarines. In contrast, now new applications spur from environmental applications, such as the detection of whales and other sea creatures [10]. The paper discusses Capon beamforming (after the seminal work [11]) introducing a new robust approach. Indeed, since very often no accurate prior information of the direction of arrival of the target is available, Capon beamforming might suffer from performance degradation due to the steering vector error dominated by large DoA mismatch. The new approach can mitigate such an issue as shown in the experimental results.

Funding: This research was funded by MIUR PON SCN 00393 "Safety & Security System for Sea Enviroment (S4E)".

Institutional Review Board Statement: Not applicable.

Informed Consent Statement: Not applicable.

Data Availability Statement: No new data were created or analyzed in this study.

Conflicts of Interest: The authors declare no conflict of interest.

References

1. Leadership Council of the Sustainable Development Solutions Network. Indicators and a Monitoring Framework for the Sustainable Development Goals. 2015. Available online: https://sustainabledevelopment.un.org/content/documents/2013150612-FINAL-SDSN-Indicator-Report1.pdf (accessed on 25 November 2020).
2. Pieri, G.; Cocco, M.; Salvetti, O. A marine information system for environmental monitoring: ARGO-MIS. *J. Mar. Sci. Eng.* **2018**, *6*, 15. [CrossRef]
3. Moroni, D.; Pieri, G.; Tampucci, M. Environmental decision support systems for monitoring small scale oil spills: Existing solutions, best practices and current challenges. *J. Mar. Sci. Eng.* **2019**, *7*, 19. [CrossRef]
4. Hu, K.; Zhang, Y.; Lu, F.; Deng, Z.; Liu, Y. An Underwater Image Enhancement Algorithm Based on MSR Parameter Optimization. *J. Mar. Sci. Eng.* **2020**, *8*, 741. [CrossRef]
5. Vargas, D.; Jayaratne, R.; Mendoza, E.; Silva, R. On the Estimation of the Surface Elevation of Regular and Irregular Waves Using the Velocity Field of Bubbles. *J. Mar. Sci. Eng.* **2020**, *8*, 88. [CrossRef]
6. Selimović, D.; Lerga, J.; Prpić-Oršić, J.; Kenji, S. Improving the Performance of Dynamic Ship Positioning Systems: A Review of Filtering and Estimation Techniques. *J. Mar. Sci. Eng.* **2020**, *8*, 234. [CrossRef]
7. D'Acunto, M.; Moroni, D.; Puntoni, A.; Salvetti, O. Optimized Dislocation of Mobile Sensor Networks on Large Marine Environments Using Voronoi Partitions. *J. Mar. Sci. Eng.* **2020**, *8*, 132. [CrossRef]
8. Xiong, M.; Zhuo, J.; Dong, Y.; Jing, X. A Layout Strategy for Distributed Barrage Jamming against Underwater Acoustic Sensor Networks. *J. Mar. Sci. Eng.* **2020**, *8*, 252. [CrossRef]
9. Hao, Y.; Zou, N.; Liang, G. Robust capon beamforming against steering vector error dominated by large direction-of-arrival mismatch for passive sonar. *J. Mar. Sci. Eng.* **2019**, *7*, 80. [CrossRef]
10. Maranda, B.H. Passive sonar. In *Handbook of Signal Processing in Acoustics*; Springer: Berlin/Heidelberg, Germany, 2008; pp. 1757–1781.
11. Capon, J. High-resolution frequency-wavenumber spectrum analysis. *Proc. IEEE* **1969**, *57*, 1408–1418. [CrossRef]

Journal of
Marine Science and Engineering

Article

An Underwater Image Enhancement Algorithm Based on MSR Parameter Optimization

Kai Hu [1,2,*], Yanwen Zhang [1,2], Feiyu Lu [1,2], Zhiliang Deng [1,2] and Yunping Liu [1,2]

[1] School of Automation, Nanjing University of Information Science & Technology, Nanjing 210044, China; zyw_wsri@163.com (Y.Z.); lu1136055775@163.com (F.L.); 002858@nuist.edu.cn (Z.D.); liuyunping@nuist.edu.cn (Y.L.)

[2] Jiangsu Collaborative Innovation Center of Atmospheric Environment and Equipment Technology (CICAEET), Nanjing University of Information Science & Technology, Nanjing 210044, China

* Correspondence: nuistpanda@163.com or 001600@nuist.edu.cn; Tel.: +86-137-7056-9871

Received: 1 September 2020; Accepted: 22 September 2020; Published: 25 September 2020

Abstract: The quality of underwater images is often affected by the absorption of light and the scattering and diffusion of floating objects. Therefore, underwater image enhancement algorithms have been widely studied. In this area, algorithms based on Multi-Scale Retinex (MSR) represent an important research direction. Although the visual quality of underwater images can be improved to some extent, the enhancement effect is not good due to the fact that the parameters of these algorithms cannot adapt to different underwater environments. To solve this problem, based on classical MSR, we propose an underwater image enhancement optimization (MSR-PO) algorithm which uses the non-reference image quality assessment (NR-IQA) index as the optimization index. First of all, in a large number of experiments, we choose the Natural Image Quality Evaluator (NIQE) as the NR-IQA index and determine the appropriate parameters in MSR as the optimization object. Then, we use the Gravitational Search Algorithm (GSA) to optimize the underwater image enhancement algorithm based on MSR and the NIQE index. The experimental results show that this algorithm has an excellent adaptive ability to environmental changes.

Keywords: underwater image enhancement; Multi-Scale Retinex; parameter optimization; no reference image quality assessment

1. Introduction

Marine resources are abundant, but the marine environment is complicated, and underwater exploration by humans is difficult. Therefore, underwater robots [1,2] are widely used in the exploration of marine resources and for underwater target recognition [3–7]. However, due to the absorption of light and the scattering and diffusion of floating objects, underwater images often have distortion, color deviation, low contrast and low definition.

There are many existing non-underwater image enhancement algorithms, such as Histogram Equalization [8] (HE). HE transforms the original image into an output image with roughly the same number of pixels at most gray levels through pixel value transformation. This algorithm has a simple structure and low computation, but HE is not good at improving the local contrast of the image. In order to improve the deficiency of HE, Pizer et al. [9] proposed an Adaptive Histogram Equalization algorithm (AHE)—AHE redistributes the brightness value of the image by calculating the histogram of each significant area of the image, making it more suitable for improving the local contrast of the image. However, the AHE algorithm has other problems, such as amplified noise. In order to solve the shortcomings of the AHE algorithm, Zuiderveld [10] proposed a Contrast Limit Adaptive Histogram Equalization (CLAHE) algorithm. In addition, algorithms such as color-corrected white balance [11] and gray-scale edge assumption [12] have also been developed. However, using the existing image enhancement algorithms directly in underwater environments may cause the problem of limited enhancement or excessive enhancement. These algorithms can only deal with the contrast of the image to a certain extent, but they cannot solve the problem of the color shifting of underwater images; it is even possible for the algorithms to cause the image to have low local contrast and distortion.

Edwin H. Land proposed the Retinex theory [13]. The Retinex algorithm can balance the three aspects of edge enhancement, dynamic range compression and color constants in the image processing process, but when the light intensity is large, the problem of exposure will occur, resulting in the phenomenon of low local contrast in the image. Zhang et al. [14] proposed the extension of the multi-scale Retinex underwater image enhancement algorithm and the Multi-Scale Retinal with Color Recovery (MSRCR) algorithm [15] to the CIELAB color space, which successfully suppressed the halo phenomenon in the process of image enhancement; however, the algorithm has many parameters, so the calculation speed is slow. Lei Fei et al. extracted the color–grayscale ratio of RGB underwater images and carried out the tensile interception of Multi-Scale Retinex (MSR) [16] enhanced images; then, they used the grayscale ratio to restore underwater images, which improved the color distortion of underwater images and enhanced their clarity. However, in an image with a large local luminance difference, the MSR-enhanced image has a halo, resulting in the phenomenon that the local contrast is too low in the image. Since then, scholars have continued to improve the Retinex algorithm [17–20].

There are two other important research directions in this field: algorithms based on the Dark Channel Prior (DCP) algorithm and algorithms based on the Generative Adversarial Networks (GAN). He et al. proposed the DCP [21] algorithm, which has been widely used in the field of image enhancement [22,23] due to its simple defogging model and good defogging effect. Zhu et al. proposed unpaired image-to-image translation algorithm using cycle-consistent adversarial networks [24]. Image enhancement algorithm based on GaN is currently a very active research area.

Underwater image enhancement algorithm improves the stability of underwater robot in various application. Zhao et al. [25] applied their improved underwater image enhancement algorithm to Autonomous Underwater Vehicle (AUV). The experimental result shows that the AUV visual SLAM can achieve accurate positioning results by underwater image enhancement algorithm. Wei et al. [26] applied the improved Retinex algorithm to the Underwater Vehicle Manipulator System (UVMS) to track and capture underwater objects. The experimental result shows that the algorithm can improve the precision of AUV grasping.

In general, the existing underwater image enhancement algorithms use fixed parameters. Although they can improve the visual quality of underwater images to some extent, there are also problems of amplified noise and color distortion. At the same time, the existing MSR-based underwater image enhancement algorithms may be particularly effective when dealing with a specific underwater environment, and the existing underwater image enhancement algorithms are unstable when facing different underwater environments. In the traditional MSR algorithm, a fixed gaussian weighting factor is selected. If the image is taken from a cloudy water body, the enhanced image will have color bias and distortion. As shown in Figures 20c and 21c, the effect is unstable. The traditional DCP algorithm uses the water environment color A as the point with the brightness of the top 0.1% in the dark channel image. If the image is taken from the water with strong illumination, the image enhancement effect will be poor, and the color of the water environment will be left in the enhanced image. As shown in Figures 22d and 23d, the effect is unstable. Therefore, it is important to determine the appropriate parameters according to the changes of different scenarios.

At the same time, the actual computing conditions of underwater robots are limited, and it is difficult to meet the real-time requirements of image processing. Therefore, this paper attempts to reduce the computational complexity of the algorithm and thus to improve its ability to be used in engineering applications.

The structure of this article is as follows—Section 1 introduces the development of underwater image enhancement work; Section 2 briefly introduces the idea of this article and the purpose of each workflow; Section 3 briefly introduces relevant background knowledge; Section 4 introduces the selection method of the parameters and the no reference image quality assessment index; Section 5 introduces the main work content of this article; Section 6 conducts experiments on the algorithm proposed in this article and objectively analyzes its performance; and Section 7 presents the conclusion of the article.

2. Methodology

The existing underwater image enhancement algorithms use fixed parameters, which can improve the visual quality of underwater images to some extent, but the existing underwater image enhancement algorithms are unstable when facing different underwater environments. MSR-PO is based on the MSR algorithm, using non-reference image quality assessment indexes to optimize the parameters to achieve a stable underwater image enhancement effect and improve the efficiency of underwater exploration by underwater robots. Firstly, we find a suitable non-reference image quality assessment index for unreferenced images. Secondly, we use the traversal method to select and optimize the parameters. Finally, in order to improve the enhancement speed, we use the Gravitational Search Algorithm (GSA) [27] to replace the traversal method for parameter optimization and propose a strategy for optimization only in stable frames to increase the speed of the enhancement.

2.1. Selection of a Non-Reference Image Quality Assessment Index

We need to find a suitable non-reference image quality assessment index. There are many no reference image quality assessment indexes, and the most suitable one needs to be selected. In this paper, Gaussian noise of different degrees is added to the image to observe the changes of multiple classic no reference image quality assessment indexes. If the changes of a no reference image quality assessment index could correctly reflect the change trend of noise, and its dynamic range was large, the indicator would be selected and applied in this article.

2.2. Selection of Target Parameters to Be Optimized in MSR

We need to use the traversal method to find the appropriate algorithm parameters. The main method is to traverse the parameters in the MSR and observe the dynamic ranges of the no reference image quality assessment indexes. Generally speaking, the parameters corresponding to the indexes with large dynamic ranges can meet the requirements of the tasks in this paper. On this basis, if the indicator changes linearly, the optimal parameters could be directly calculated, and if the indicator changes nonlinearly, optimization algorithms would need to be used for optimization.

In specific applications, the computing power of an underwater robot is limited, the real-time requirement for image enhancement is high and parameter screening is required to reduce the amount of calculation. The main method used is to calculate a set of no reference image quality assessment index data after each parameter traversal and calculate the influence of the data variance comparison parameters. If the influence of a parameter increases with the variance, the parameter is retained; otherwise, it is deleted.

2.3. Improve Engineering Application Capabilities

The traversal method is stable but slow. This paper attempts to optimize the parameters by using the GSA optimization algorithm and conducts multiple sets of comparative experiments to ensure the stability of the GSA algorithm.

When enhancing video frames, considering that underwater robots require real-time image enhancement capabilities, this article attempts to propose a strategy that only optimizes the frame to increase the enhancement speed; the feasibility of this strategy is verified in this work.

The work flow chart of the algorithm in this paper is shown in Figure 1.

When the existing underwater image enhancement algorithm faces different underwater environments, the enhancement effect is unstable due to the limitation of the algorithm's own fixed parameters. To solve these problems, we use the non-reference image quality evaluation index as the optimization index to optimize the parameters in the MSR algorithm and improve the enhancement speed. Based on this, we propose MSR-PO algorithm.

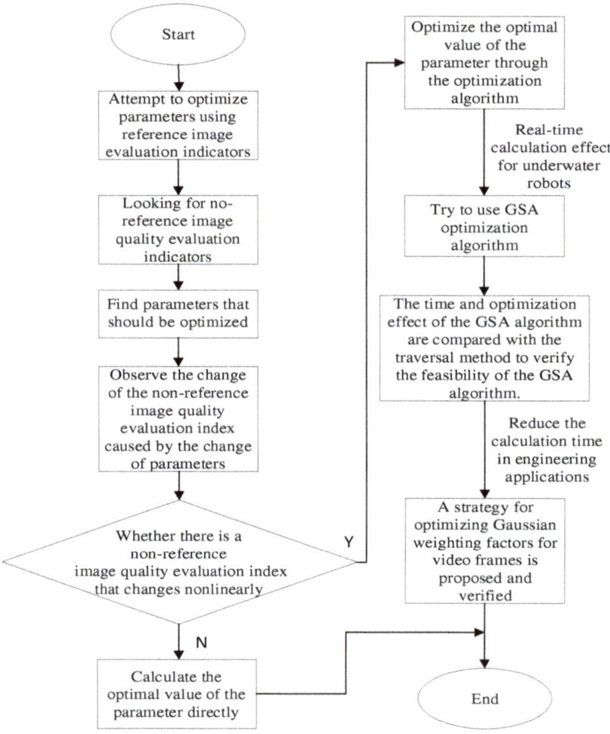

Figure 1. The working idea of this paper.

3. Background Knowledge

3.1. Retinex Algorithm Introduction

3.1.1. Single Scale Retinex (SSR)

The single-scale Retinex algorithm works by improving the center/surround Retinex. The basic idea of the center/surround Retinex algorithm is to estimate the brightness of the center pixel by assigning different weights to the pixels surrounding the target. The proportional relationships of these weights are determined by the surrounding function. Jobson proposed a single-scale Retinex algorithm (SSR) [28] based on this theory. The basic content of the algorithm is as follows. The algorithm assumes that the image is $I(x, y)$, the brightness image is $L(x, y)$, the reflected image is $R(x, y)$; then:

$$I(x, y) = L(x, y) \cdot R(x, y). \tag{1}$$

The purpose of the algorithm is to obtain $R(x, y)$ from $I(x, y)$. The algorithm assumes that the image is smooth and approximates the low-frequency part of the image. The logarithmic domain processing makes the perception of the image more similar to the human eye's perception of brightness. The single-scale Retinex can be expressed as

$$\ln[R(x,y)] = \ln\frac{I(x,y)}{L(x,y)} = \ln[I(x,y)] - \ln[I(x,y) * G(x,y)]. \tag{2}$$

In Equation (2), $G(x,y)$ is a low-pass filter used to estimate the brightness map; subtracting the brightness map from the original image can restore the original color of the object. Jobson's theory proves that the Gaussian convolution function can process the image locally to better enhance the image; thus, $G(x,y)$ is usually a Gaussian function, and the expression is

$$G(x,y) = \lambda \cdot e^{\frac{-(x^2+y^2)}{c^2}}. \tag{3}$$

In Equation (3), c is a scale constant. The smaller the c is, the larger the dynamic compression range would be. On the contrary, the larger the c is, the more severe the image sharpness would be. λ is a constant matrix, and its value must satisfy Equation (4):

$$\iint G(x,y)\,dx\,dy = 1. \tag{4}$$

3.1.2. Multi-Scale Retinex Algorithm (MSR)

In view of the problem that the SSR algorithm has poor processing effects in terms of image detail enhancement and color fidelity, Joson et al. proposed a multi-scale Retinex algorithm (MSR).

The multi-scale Retinex algorithm is the weighted average of the single-scale Retinex algorithm. $r(x,y)$ is used to represent the final calculation result of the reflected image. The multi-scale Retinex algorithm can be expressed as

$$\ln[r(x,y)] = \sum_{i=1}^{N} w_i \left\{ \ln[I(x,y)] - \ln[I(x,y) * G_i(x,y)] \right\}. \tag{5}$$

In Equation (5), n represents the number of scales. In general, $n = 3$ is a color image and w_i is a Gaussian weighting coefficient. Assuming that the weights corresponding to each scale are equal, then $w_i = 1/3$, and $G_i(x,y)$ is a Gaussian low-pass filter:

$$G_i(x,y) = \lambda \cdot e^{\frac{-(x^2+y^2)}{c_i^2}}. \tag{6}$$

In Equation (6), c_i represents the scale.

3.2. No Reference Image Quality Assessment Index

Current underwater image quality assessment methods are divided into subjective assessment and objective assessment [29]. The subjective assessment method is mainly used for testers to observe the target image and make a subjective assessment and analysis. Generally, the Mean Opinion Score (MOS) or Differential Mean Opinion Score (DMOS) refer to the difference between the assessment scores of undistorted and distorted images by human eyes. However, in practical applications, subjective assessment is limited by many factors, and the uncertainty of subjective assessment makes it impossible to describe the method with a mathematical model; furthermore, it is difficult to achieve high-quality assessment. The objective assessment method is used to calculate the visual quality of the image through certain algorithms and mathematical methods. It includes Full Reference Image Quality Assessment (FR-IQA), Reduced Reference Image Quality Assessment (RR-IQA) and No Reference Image Quality Assessment (NR-IQA) [30]. The following is a brief introduction to four common and classic non-reference image quality assessment indexes.

3.2.1. Entropy

Entropy [31] appears in information theory to indicate the amount of information. It was not used to evaluate image quality until 1978. For the assessment of the quality of underwater images, the color information entropy assessment method is adopted. The RGB color space is selected; then, the information entropy is obtained based on the three color channels of the image, and the average value is finally obtained. The formula for information entropy is as follows:

$$H = - \sum_{0}^{255} P(i) \log_2 P(i). \tag{7}$$

In Equation (7), i represents the gray value, and $P(i)$ represents the proportion of pixels with a gray value of i to all pixels in the image. Color image entropy is the average of the information entropy of the three channels R, G and B. The greater the information entropy of the image is, the more information the image has and the better the image quality is.

3.2.2. UCIQE

The Underwater Color Image Quality Evaluation (UCIQE) [32] is a new image quality evaluation index proposed by Yang Miao et al. The objective image quality evaluation index adopts the linear combination of chroma, saturation and brightness as the measurement method of image quality. The higher the UCIQE score is, the better the image restoration would be. The UCIQE is defined as

$$UCIQE = c_1 \times \sigma_c + c_2 \times con_l + c_3 \times \mu_s. \tag{8}$$

In Equation (8), σ_c is the standard deviation of chroma, con_l is the brightness contrast, μ_s is the average value of saturation and c_1, c_2 and c_3 are the weighting coefficients. After a large number of underwater image experiments, the coefficients finally obtained under the underwater test data sets are $c_1 = 0.4680$, $c_2 = 0.2745$ and $c_3 = 0.2576$. Generally speaking, the higher the UCIQE index is, the better the quality of the underwater image would be; otherwise, poorer image quality would be obtained.

3.2.3. NIQE

The Natural Image Quality Evaluator (NIQE) [33] is an image quality evaluation index that was proposed by Mittal et al. in 2012. The algorithm is based on a probability model. In this algorithm, the image is first mean-removed normalized; then, the processed image is fitted by generalized Gaussian distribution and a multivariate Gaussian model to obtain the parameters. Finally, the distance between the parameters and the corresponding parameters in the image library is the score of the image. The NIQE algorithm calculates the distance between the distorted image and the original image in the image library. This distance is used to measure the image quality. If the distance is small, this means that the image is similar to the natural image in the image library, the degree of distortion is small and the image quality is good. Conversely, if the distance between the distorted image and the original image in the image library is large, it means the degree of distortion is large and the image quality is poor.

3.2.4. UIQM

Karen et al. [34] proposed an Underwater Image Quality Measurement (UIQM) based on the characteristics of underwater image degradation. The UIQM method uses the linear combination of the underwater image colorfulness measurement (UICM), the underwater image sharpness measurement (UISM) and the underwater image contrast measurement (UIConM) as the evaluation basis.

$$UIQM = c_1 \times UICM + c_2 \times UISM + c_3 \times UIConM. \tag{9}$$

In Equation (9), c_1, c_2 and c_3 are the weighting factors of the measurement components in the linear combination. The method shows that the selection of weighting factors should be based on specific quality evaluation requirements. When the color quality requirements for underwater images are high, the value assigned to c_1 is large; when the image contrast and clarity quality requirements are high, the weight values assigned to UISM and UIConM need to be large, and the corresponding c_2 and c_3 values are large. In this paper, the weighting factors are $c_1 = 0.0282$, $c_2 = 0.2953$ and $c_3 = 3.5753$.

4. Selection of Parameters and Assessment Indexes

Whether the no reference image quality assessment index can be used for optimization depends on whether the index is able to correctly reflect the changing trend of noise and to have a large dynamic range. By adding different degrees of Gaussian noise to the image, this paper evaluates the image quality with four kinds of no reference image quality assessment indexes, including information entropy, UCIQE, NIQE and UIQM. The next step is to observe the change trends and dynamic ranges of these four indexes and choose a no reference image quality assessment index which can optimize the algorithm. Then, we use the index to traverse the parameters, observe the dynamic range of the NIQE index and select the parameters that can be optimized. In order to increase the speed of the calculation, the parameters need to be screened again, and the parameters with a larger influence must be selected.

4.1. Selection of Assessment Indexes

In this paper, we intend to optimize the parameters by using NR-IQA to achieve stable underwater image enhancement and improve the efficiency of underwater exploration by underwater robots. The selection of the no reference image quality assessment index mainly includes two aspects:

(1) The change of the no reference image quality assessment index correctly reflects the change of the quality of underwater images;
(2) The dynamic range of the non-reference image quality assessment index is large.

The main method used is to select four representative images (Figure 2) that are commonly used in the field of underwater image quality enhancement as experimental objects. Figure 2a is a greenish image of large targets. Figure 2b is a greenish image with light spot. Figure 2c is a bluish image of large targets. Figure 2d is a bluish image of small targets. These four images are selected in this paper to simulate different underwater environments. Gaussian noise is added to the images, and the variance in the Gaussian noise function is from 0 to 0.1 in steps of 0.002. With the increase of Gaussian noise in the image, we need to calculate four classic non-reference image quality assessment indexes and count their change trends to obtain four graphs as shown in Figure 3, where the X axis is the variance of the added Gaussian noise and the Y axis is the calculated four types of non-reference image quality assessment indexes. In Figure 3, the results of picture 1 are shown in Figure 2a, the results of picture 2 are shown in Figure 2b, the results of picture 3 are shown in Figure 2c, and the results of picture 4 are shown in Figure 2d.

Analyzing Figures 2 and 3, we find that, with the increase of variance in Figure 3a, the UCIQE indicator shows an upward trend, indicating that the image quality is improving, but the UCIQE indicator ought to increase as the image quality becomes better, which is contrary to the requirement of this article. As the variance increases in Figure 3b,d, the UIQM indicator and information entropy show a trend of rising first and then decreasing, indicating that the image quality first improves and then becomess worse, so they are unable to correctly reflect the changing trends of the image quality, which is inconsistent with the requirements of this paper. As the variance increases in Figure 3c, the NIQE shows an increasing trend. The larger the NIQE is, the bigger the difference is, meaning that the image quality is getting worse. The NIQE indicator correctly reflects the trend of image quality change; the difference between the minimum and maximum of the NIQE index can reach about 13, indicating that its dynamic range changes greatly; thus, this article chooses the NIQE indicator as the optimization indicator.

(**a**) Picture1: a greenish sample of large targets

(**b**) Picture2: a greenish sample with light spot

(**c**) Picture3: a bluish sample of large targets

(**d**) Picture4: a bluish sample of small targets

Figure 2. Underwater test pictures.

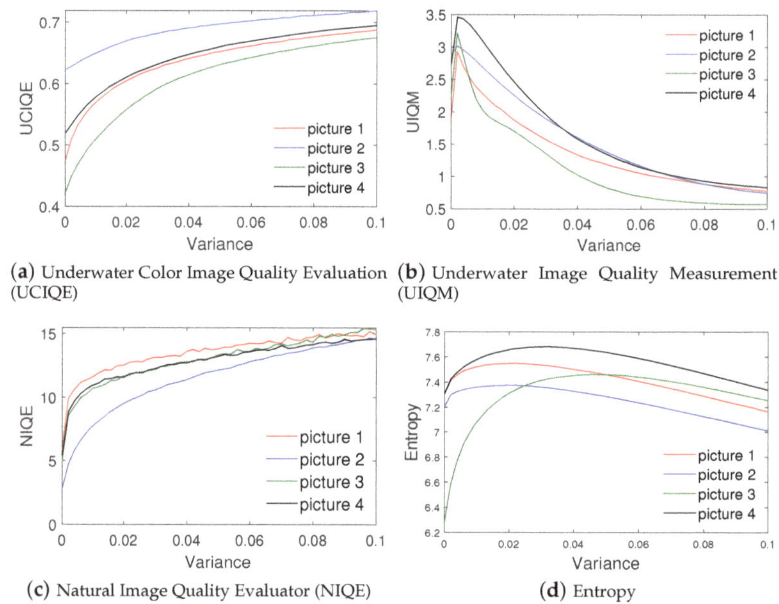

(**a**) Underwater Color Image Quality Evaluation (UCIQE)

(**b**) Underwater Image Quality Measurement (UIQM)

(**c**) Natural Image Quality Evaluator (NIQE)

(**d**) Entropy

Figure 3. The degree of response of the four indexes to Gaussian noise.

In the specific calculation, NIQE first normalizes the image pixels, extracts the features of the salient image blocks, and then performs image quality calculation. By comparing the natural image and the distorted image, the mean v_1, v_2 and variance matrices σ_1, σ_2 of natural and distorted images are obtained, and finally the image quality is measured by calculating the distance between the natural image and distorted image fitting parameters; the calculation formula is as follows:

$$D(v_1, v_2, \sigma_1, \sigma_2) = \sqrt{(v_1 - v_2)^T (\frac{\sigma_1 + \sigma_2}{2})^{-1} (v_1 - v_2)}. \tag{10}$$

Generally speaking, the smaller the NIQE value, the better the image quality.

4.2. Selection of Optimization Parameters

In this paper, we try to optimize the parameters by using no reference image assessment indexes to achieve stable underwater image enhancement. The changes of different parameters have different effects on the MSR algorithm results, meaning that the task presented in this section is to find the parameters that have the greatest influence on the MSR results.

There are many parameters in the MSR algorithm that are set by human experience. In this paper, two sets of parameters that have relatively large influences on the MSR algorithm results are obtained through preliminary screening. The first group is the scale parameter values sigma1, sigma2 and sigma3, and the second group is the Gaussian weighting factors W1, W2 and W3, and the pre-screening methods are the same as the methods in the section above; thus, the details will not be repeated here. At present, all MSR parameter values are artificially summarized and cannot be adapted to different environmental conditions. This paper uses no reference image quality assessment indexes to optimize the parameters, but the extent to which the changes in the above parameters could affect the quality index of the no reference image is not known.

Generally speaking, if the change of a parameter can lead to a large change in the no reference image quality assessment index, then the optimization effect of this parameter will be good, which is also true for the optimization index of our work in this section.

The process of selecting optimization parameters first traverses each parameter and observes the changes of NIQE indexes. In the traversal process, the range of parameter traversal must first be determined. The range of the traversal is mainly determined by the general values of the three parameters, and the values of the three scale parameters should at least be twice as large from the first to the third parameter. Based on this, the sigma1 traversal range is set to 10–40; the sigma2 traversal range is set to 30–125; and the sigma3 traversal range is set to 160–300. In this section, we randomly select two pictures in the sea urchin database of the National Fund of China (NFC) (Figure 4) and traverse their sigma1, sigma2 and sigma3. The results are shown in Figure 5.

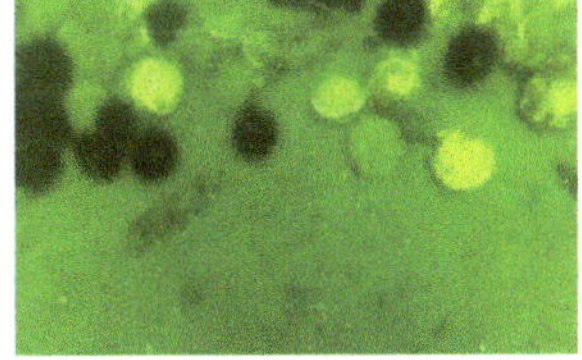

(**a**) Randomly select sample 1 from the sea urchin database of the NFC (**b**) Randomly select sample 2 from the sea urchin database of the NFC

Figure 4. Underwater test images.

In the graph of Figure 5, the abscissa is the traversal parameter, and the ordinate is the NIQE value. In Figure 5a,b, when traversing sigma1, the difference between the maximum value and the minimum value of the dynamic range of NIQE value is about 35%; in Figure 5c,d, when traversing sigma2, the difference between the maximum value and the minimum value of the dynamic range of NIQE is about 4–6%; and in Figure 5e,f, when traversing sigma3, the difference between the maximum value and the minimum value of the dynamic range of NIQE is about 1–2%.

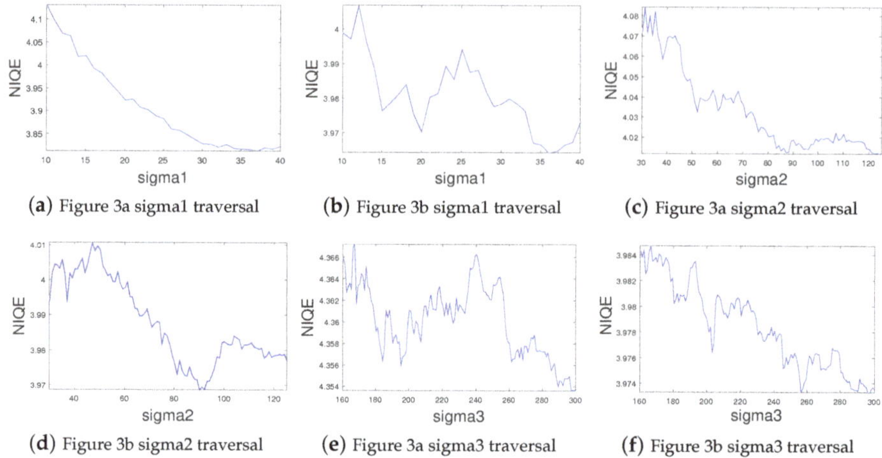

(a) Figure 3a sigma1 traversal   **(b)** Figure 3b sigma1 traversal   **(c)** Figure 3a sigma2 traversal

(d) Figure 3b sigma2 traversal   **(e)** Figure 3a sigma3 traversal   **(f)** Figure 3b sigma3 traversal

Figure 5. The result of traversing sigma1, sigma2 and sigma3 of the two pictures in Figure 4.

The second set of weight parameters has the relationship of W1 + W2 + W3 = 1. The three parameters restrict each other. The third weight W3 can be expressed as 1 − W1 − W3, so two parameters are selected for simultaneous optimization. In this article, W1 and W2 are selected for simultaneous traversal, and the traversal range of both W1 and W2 is set to 0–1. This method traverses W1 and W2 in the two pictures in Figure 4 to observe the experimental results.

As shown in Figure 6, when traversing W1 and W2, the maximum value and the minimum value of the dynamic range of NIQE can differ by about 0.6.

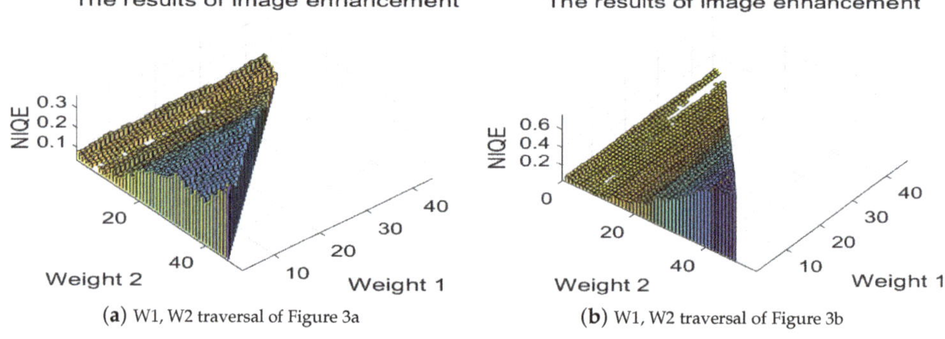

(a) W1, W2 traversal of Figure 3a   **(b)** W1, W2 traversal of Figure 3b

Figure 6. W1, W2 traversal results.

According to the above analysis, the dynamic range in the graph becomes larger as the parameter values vary, so the five parameters can be optimized. There are multiple peaks and troughs in Figures 5 and 6, indicating nonlinear changes, so the optimal values of parameters cannot be directly calculated, and an optimization algorithm is needed to obtain the optimal values of parameters.

From the above experiments, we can find that all five parameters can be optimized, and optimizing these five parameters will consume a large amount of time. This paper attempts to traverse each parameter and bring the results into the MSR algorithm for enhancement; five sets of no reference image quality assessment indexes could be obtained through the calculation, and the variance of these five sets of data could also be obtained. The greater the variance, the greater the influence of the parameter. The less influential parameters would not be optimized. Among them, the three parameters of the first group of scale parameters do not restrict each other, while the second group of Gaussian weighting factors restrict each other, and the mutually restricted parameters cannot exclude any of the factors. Therefore, this paper calculates the influence of the first group of parameters, as shown in Figure 7 and randomly selects four images from the sea urchin database of the China National Fund Commission for testing.

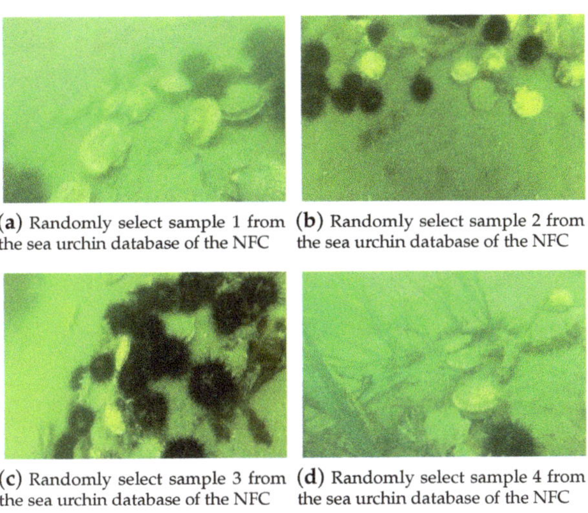

(**a**) Randomly select sample 1 from the sea urchin database of the NFC   (**b**) Randomly select sample 2 from the sea urchin database of the NFC

(**c**) Randomly select sample 3 from the sea urchin database of the NFC   (**d**) Randomly select sample 4 from the sea urchin database of the NFC

Figure 7. Underwater test images.

As shown in Figure 8, a histogram of variances with three Gaussian weighting factors for four pictures is listed. The three parameter variances are sorted as sigma1, sigma2 and sigma3 from largest to smallest. The small-scale parameter sigma1 has a greater impact on the image quality evaluation results than the mesoscale parameter sigma2 and the large-scale parameter sigma3. Therefore, sigma1 is selected for optimization in this paper.

It should be pointed out that the better the parameters, the more the indexes will be improved. With the improvement of underwater robot computing resources, we can optimize all parameters. This article only uses sigma1 as an optimization parameter and puts forward the idea of the algorithm in this paper. It is hoped that the stability can be improved at first and that the accuracy can be improved to some extent without limiting the number of optimized parameters, which could be increased later.

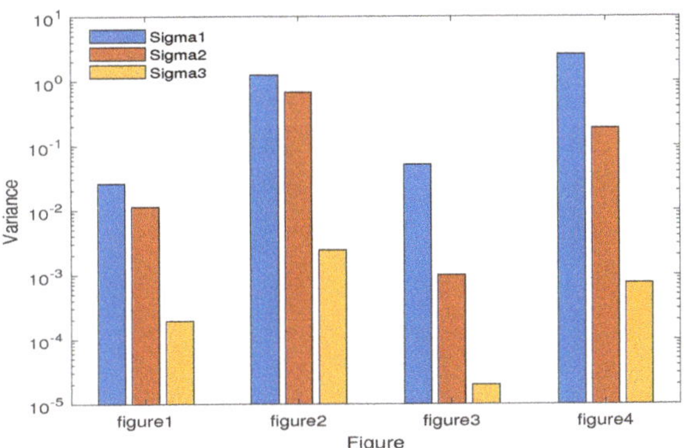

Figure 8. Parameter influence.

5. Our Algorithm

The three scales—sigma1, sigma2 and sigma3—of the existing MSR algorithm are derived from human experience. In theory, sigma3 depends on the image size while sigma1 depends on the size of target objects and the water turbidity. From the above experiments, it has been verified that sigma1, sigma2 and sigma3 have obvious impacts on image quality, and sigma1 has the greatest impact on image quality. This article makes adaptive adjustments to sigma1 of the MSR algorithm and optimizes and confirms the optimal value of sigma1 through the no reference image quality assessment index.

The existing MSR algorithm artificially sets the three Gaussian weighting factors to one-third, but this does not meet the data distribution characteristics of the underwater image color space and causes image distortion problems. It has been verified from the above experiments that W1, W2 and W3 have a great influence on image quality, and so this article makes adaptive adjustments to the Gaussian weighting factors of the MSR algorithm, and optimizes and confirms the optimal values of Gaussian weighting factors through no reference image quality assessment indexes.

In order to reduce the dimension of the optimization space and improve the optimization speed, the algorithm splits the original target (NIQE) and three variables (sigma1, w1, w2) into two steps for optimization. The first step is the optimization of the target (NIQE) and one variable (sigma1) to find the optimal sigma1. The second step is the optimization of the target (NIQE) and two variables (w1, w2) to find the optimal w1 and w2.

For practical reasons, we make an assumption: in continuous frames in dozens, the environment of the underwater Image is basically unchanged; that is, the size of the target object and the turbidity of the water body are unchanged. Then, we can adopt the strategy of optimizing only in stable frames. During this time, we do not need to repeatedly optimize the values of sigma1, w1 and w2; we can use the same parameters to perform the restoration work, which can increase the enhancement speed.

On this basis, the algorithm of this paper first obtains the acceleration of each frame image through the IMU in the underwater robot. Taking N video frames as a group, the frame with the smallest acceleration in a group of video frames is selected as the stable frame, and the optimal sigma1, first_W1 and first_W2 found by the GSA in the stable frame are used, and the remaining N-1 frame videos use sigma1, first_W1 and first_W2 to implement the image restoration.

Single image enhancement steps are shown in the workflow in Algorithm 1, and the video frame image enhancement steps are shown in Algorithm 2.

Algorithm 1 Single image enhancement steps

Input: Single image
Output: Enhanced image

    Step1. Adaptive adjustment is made for sigma1 of the MSR algorithm. Through no reference image quality assessment indexes, the GSA algorithm is used to optimize and confirm the optimal sigma1 value.

    Step2. Adaptive adjustment is made for the gaussian weighting factor of the MSR algorithm, the optimal values of w1, w2, and w3 are optimized and confirmed through no reference image quality assessment indexes.

    Step3. The optimal sigma1, w1, w2, and w3 are applied to the MSR algorithm for underwater image enhancement.

Algorithm 2 Video frame image enhancement steps

Input: A set of video frame images

Output: A set of enhanced video frame images

Step1. The underwater robot calculates the frame with the smallest acceleration in a group of video frame images through the IMU, and sets the frame with the smallest acceleration as the stable frame.

Step2. Adaptive adjustment is made for sigma1 of the MSR algorithm. Through no reference image quality assessment indexes, the GSA algorithm is used to optimize the optimal sigma1 value of the stable frame.

Step3. The optimal sigma1 value of the stable frame is used as the sigma1 value of the group of video frames.

Step4. The gaussian weighting factors optimized by the GSA algorithm for the first frame of each group are first_W1, first_W2, and the following N-1 frame videos use first_W1, first_W2 to realize the restoration of underwater images.

---

6. Experimental Analysis

6.1. Gsa Optimization Effect

Considering that the traversal method takes a long time, and the underwater environment is complex and changes greatly, in order to reduce the calculation time, we try to use the GSA optimization algorithm instead of the traversal method. Figure 9 is a time comparison diagram of the optimization of two Gaussian weighting factors in 16 video frame images using the traversal method and the GSA algorithm. From Figure 9, it can be found that the optimization efficiency of the GSA algorithm is far ahead in the two-parameter optimization problem, which is five times faster than the traversal method. The optimization algorithm is more efficient than the traversal method, but the accuracy is slightly reduced. This is well. known, and this article only makes a brief description of this fact here.

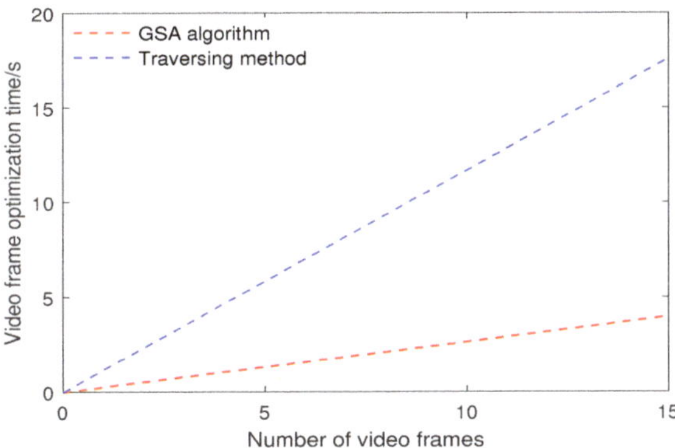

Figure 9. Gravitational Search Algorithm (GSA) and traversal method time comparison chart.

Four images (Figure 10) are randomly selected from the underwater sea urchin database and optimized to obtain Figure 11. The abscissa in Figure 11 is the number of iterations, and the ordinate represents the sum of R, G and B three-channel NIQE evaluation values. As the number of iterations increases, the score of image quality converges continuously. Figure 12 shows the optimal sigma1 found by the traversal method and the GSA algorithm. The optimal solutions found by these two methods differ in less than two scales, meaning that the GSA algorithm is feasible to use. In Figure 13, the optimization speed of the GSA algorithm is less than half of the traversal method; thus, when the underwater robot performs image enhancement, the GSA algorithm can improve the optimization speed.

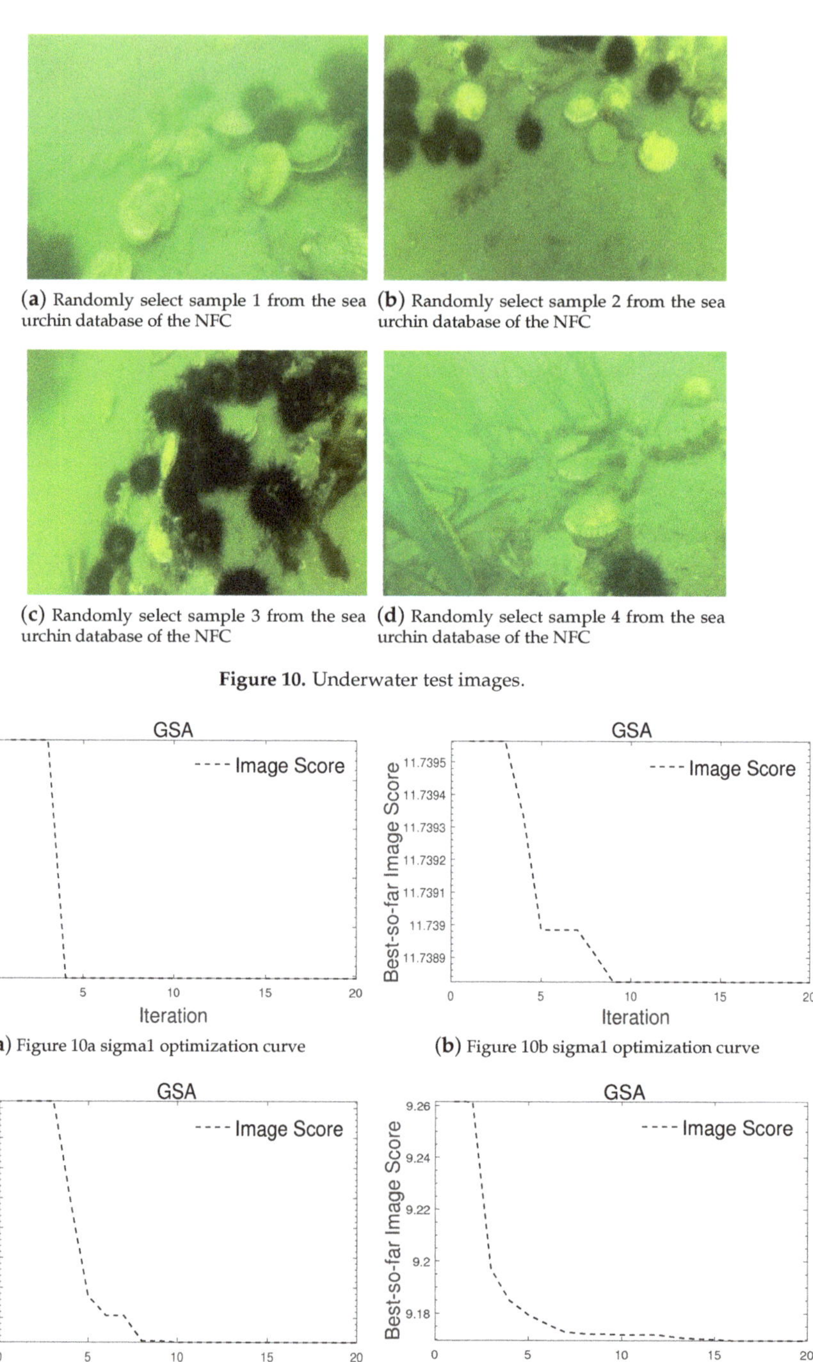

(**a**) Randomly select sample 1 from the sea urchin database of the NFC

(**b**) Randomly select sample 2 from the sea urchin database of the NFC

(**c**) Randomly select sample 3 from the sea urchin database of the NFC

(**d**) Randomly select sample 4 from the sea urchin database of the NFC

Figure 10. Underwater test images.

(**a**) Figure 10a sigma1 optimization curve

(**b**) Figure 10b sigma1 optimization curve

(**c**) Figure 10c sigma1 optimization curve

(**d**) Figure 10d sigma1 optimization curve

Figure 11. Sigma1 optimization fitness evolution curve.

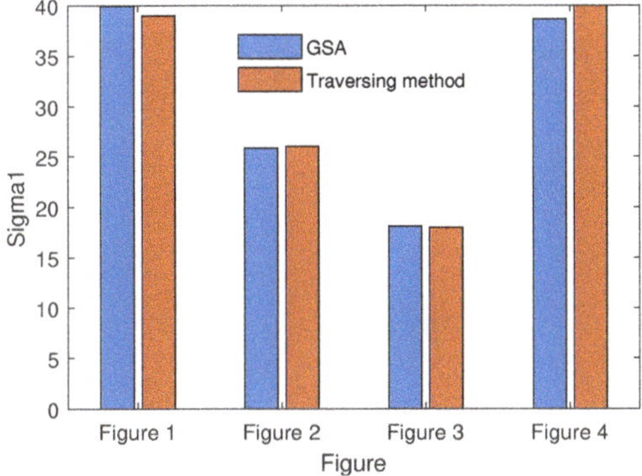

Figure 12. Comparison of the optimal values of optimization and traversal.

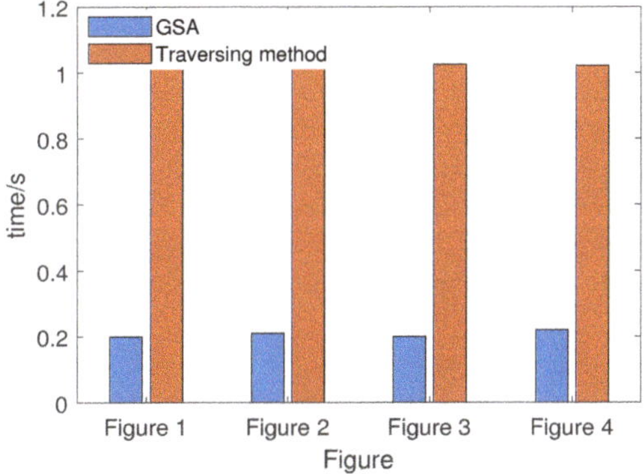

Figure 13. Comparison of the optimal values of optimization and traversal.

6.2. Strategy and Time Analysis of Optimizing Only in Stable Frames

Figure 14 shows a set of 10 consecutive frames of video images collected by an underwater robot. The sequence of these video frames is from left to right and from top to bottom. The subsequent optimization experiments correspond to the order of these 10 pictures.

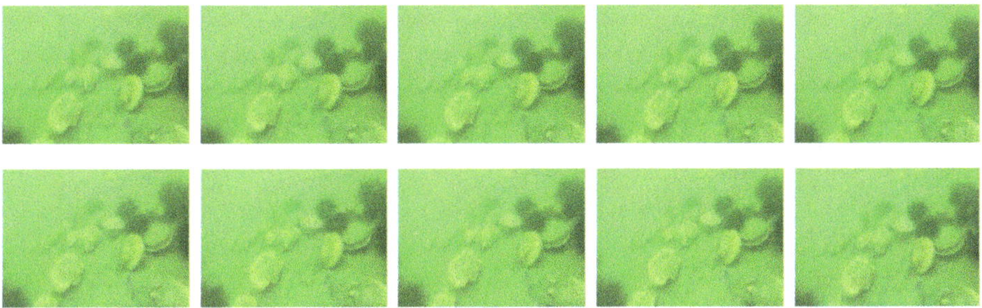

Figure 14. Experimental test images.

Figure 15 shows the iterative curves of the optimize_Score value for these 10 continuous frames of video images obtained by the GSA algorithm. The abscissa is the number of iterations, and the ordinate is the NIQE value after the RGB channel fusion of the image. As the number of iterations increases, the image quality scores continue to converge. Judging from the scores of these consecutive video frames, the overall difference is not large—the largest gap is only 0.35—and the overall curve completely converges within 20 epochs.

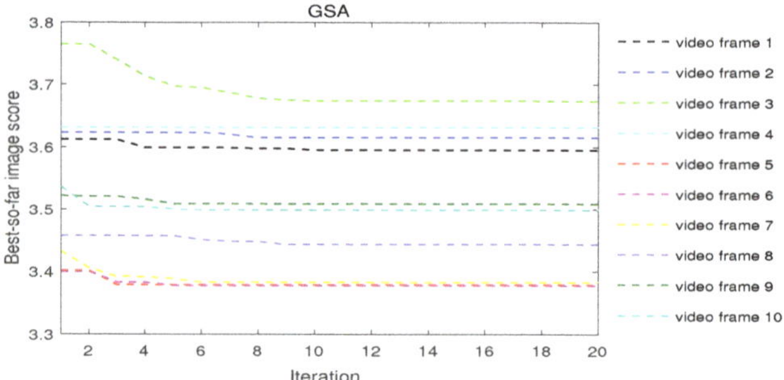

Figure 15. GSA image quality score optimization curve.

Figure 16 shows the optimal Gaussian weighting factors (best_W1 and best_W2) obtained by the traversal method and the optimized Gaussian weighting factor parameters (optimize_W1 and optimize_W2) corresponding to the GSA in Figure 15. In the Figure 16, the x-axis is the video frame order, and the y-axis is the Gaussian weighting factor value. From Figure 16, we can se that the best_W1 and optimize_W1 of the 10-frame video are concentrated between 0.1–0.2, the best_W2 and optimize_W2 are basically concentrated between 0.3–0.4, and the Gaussian weighting factors in the same scene are stable.

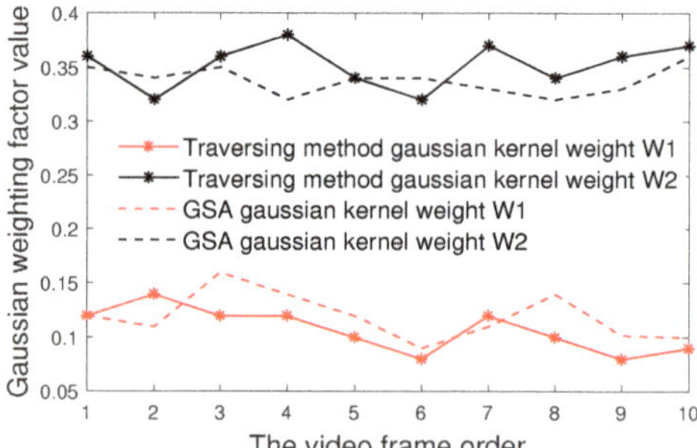

Figure 16. Traversal method and GSA algorithm used to optimize Gaussian weighting factors.

As can be seen in Figure 16, for w1 and w2, these two parameters show basically no change in the 10 frames of pictures, so the assumption in Section 5 can be confirmed.In continuous frames in dozens, the underwater image environment basically remains unchanged; that is, the size of the target object and the turbidity of the water body are unchanged. Then, we can adopt the strategy of optimization only in stable frames. During this time, we do not need to repeatedly optimize the values of sigma1, w1 and w2. We can use the same parameters to perform restoration work, which can increase the enhancement speed.

It should be noted that, through experiments, we find that not only are the 10 continuous frames as listed in this article are basically unchanged, but also in hundreds of consecutive frames, the image parameters remain basically unchanged; that is, these 10 frames can be replaced with hundreds of frames.

As shown in the Figure 17, the underwater robot used in the experiment is a new type of underwater robot launched by Aquabotix. As shown in the Figure 18, the robot operating system used in the experiment is configured with the GPU hardware platform of Nvidia Geforce GTX1060Ti. Figure 19 is a graph of the running time of the video frame Gaussian weighting factor optimization strategy, where the x-axis represents the number of video frames and the y-axis is the time for which the algorithm runs. The GSA algorithm needs 205 ms to optimize Gaussian factors, and the subsequent image color restoration is about 5 ms for each image.

Figure 17. Autonomous underwater vehicle.

Figure 18. Autonomous underwater vehicle operating system.

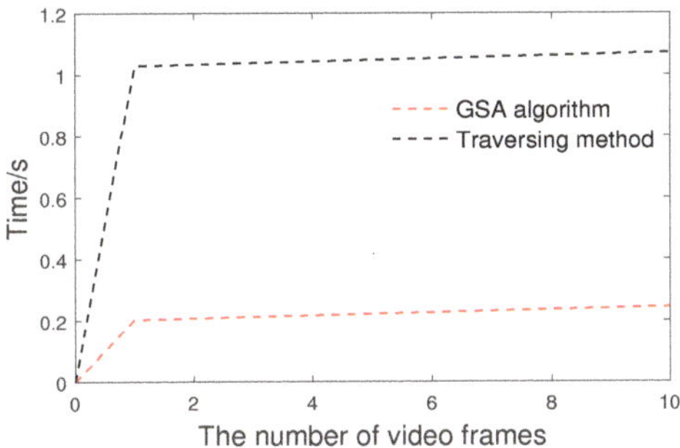

Figure 19. Video frame Gaussian weighting factor optimization time curve.

In terms of time, a stable working time TT1 in this paper is shown in Equation (11):

$$TT1 = T1 + 10 * T2, \tag{11}$$

where TT1 is the consumed time of this algorithm in 10 consecutive frames, including one optimization time of T1, and the restoration time T2 of an image after obtaining optimization parameters 10 times. The number of shooting frames of the underwater camera is 30 FPS/s, then the time resource TT1 for this algorithm should be less than 333 ms.

By performing calculation with the GPU hardware platform of Nvidia Geforce GTX1060Ti, T1 in this article is 205 ms and T2 is 5 ms, so TT1 is 255 ms, and TT1 is less than 333 ms, which meets the calculation time requirement given in the project.

Since the optimization time T1 is longer than T2, and the image can be stable for hundreds of frames, according to our own needs, we can increase the number of frames from 10 to conserve more computing resources for further analysis and control. Due to the time limitation in engineering applications, in the MSR-OP algorithm, images with resolution of 400 × 300 can be restored and optimized in a group of 10 frames three times per second. If higher resolution is required, such as 800 × 600 images, 30 frames can be set as a group to meet the requirements of engineering applications, and the MSR-PO algorithm can be restored and optimized once per second.

6.3. Overall Algorithm Effect Analysis

In order to prove the applicability and superiority of the proposed enhancement algorithm, this paper uses the Contrast Limit Adaptive Histogram Equalization (CLAHE), the classic Multi-Scale Retinex with Color Restoration (MSRCR) algorithm and the Dark Channel Prior (DCP) algorithm for comparison. The underwater images processed by different algorithms are analyzed and compared from the two perspectives of subjective visual perception and an objective evaluation index.

In our research, hundreds of images are analyzed in the general underwater image database and the sea urchin database of the China National Fund Committee. Four groups of underwater image enhancement contrast experiments are randomly selected for demonstration here. The first (Figure 20a) underwater image is from the universal underwater database, and the next three underwater pictures (Figures 21a, 22a and 23a) are randomly selected from the sea urchin database of the National Fund of China.

Images (b–d) in Figures 20–23 are the effect pictures of different algorithms. Image (b) in Figures 20–23 is that of the CLAHE algorithm, image (c) in Figures 20–23 is that of the MSRCR algorithm, image (d) in Figures 20–23 is that of the Dark Channel Prior algorithm and image (e) in Figures 20–23 shows the image generated by MSR-PO algorithm.

Analyzing the results of these three image enhancement algorithms through Figures 20–23, it can be seen that the contrast and sharpness of the image processed by these three algorithms are improved to different extents. In scene 1 and scene 2, the CLAHE algorithm and the parameter-optimized MSR algorithm enhance the visual effect of the image very well, but in scene 3 and scene 4, the effect of the CLAHE algorithm is obviously not as good as MSR-PO algorithm. This is because the color channel of the CLAHE algorithm is limited, so the image still has the problem of color casting. In the four scenes, although the effect of the MSRCR algorithm can restore the details of the image well, the overall color of the image fades to white, and there is a slight distortion problem. In general, the MSRCR algorithm is visually inferior to the MSR-PO algorithm. In scene 3 and scene 4, the enhanced image of DCP algorithm is slightly green. This is because the DCP algorithm has a dark channel depth that is not obvious, so the image transmittance cannot be estimated, and the image contrast and color difference are not repaired well. Compared with the other two algorithms, MSR-PO algorithm performs better, and it can be widely used in different scenarios because its effect is more stable.

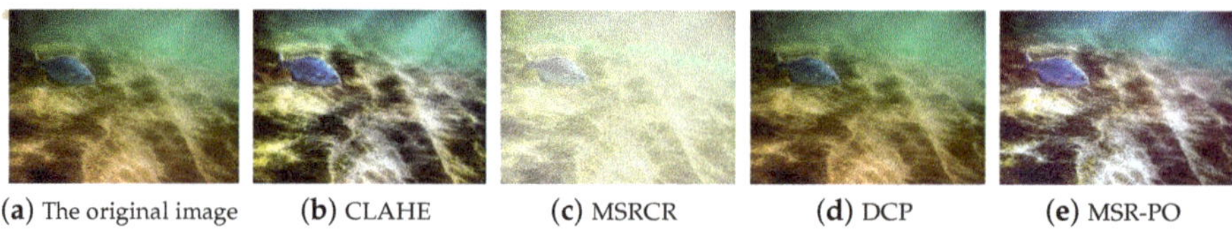

(**a**) The original image   (**b**) CLAHE   (**c**) MSRCR   (**d**) DCP   (**e**) MSR-PO

Figure 20. Scene 1 comparison.

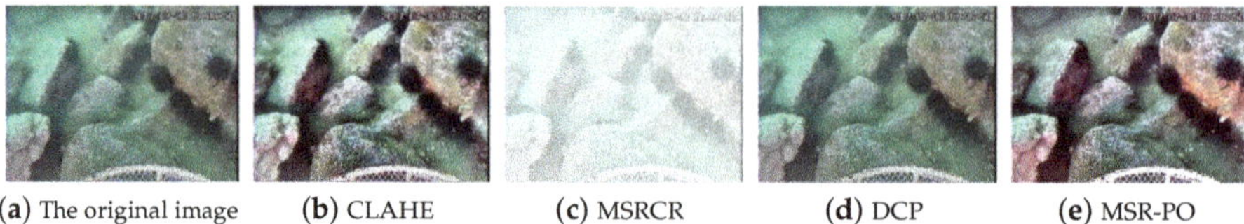

(**a**) The original image   (**b**) CLAHE   (**c**) MSRCR   (**d**) DCP   (**e**) MSR-PO

Figure 21. Scene 2 comparison.

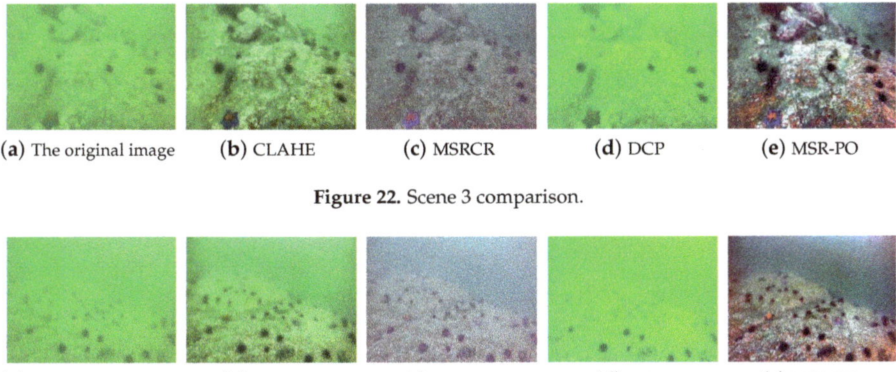

(**a**) The original image   (**b**) CLAHE   (**c**) MSRCR   (**d**) DCP   (**e**) MSR-PO

Figure 22. Scene 3 comparison.

(**a**) The original image   (**b**) CLAHE   (**c**) MSRCR   (**d**) DCP   (**e**) MSR-PO

Figure 23. Scene 4 comparison.

Table 1 uses information entropy, UCIQE and NIQE to evaluate the images in Figures 20–23. It is difficult to determine which algorithm has the best image enhancement effect only by observing Table 1. Therefore, the average index is used to prove the superiority of the MSR-PO algorithm. The final result is shown in Table 2. An index evaluates the enhanced underwater images with different algorithms, and the best results in the table are bold fonts. After the normalization operation, the following can be found: the highest average quality index of the four scenes is the MSR-PO algorithm; furthermore, the performance of the MSR-PO algorithm is more stable.

Table 1. Objective assessment index of underwater image quality.

Scene	Enhanced Algorithm	Entropy	UCIQE	NIQE
	None	7.570	0.619	2.885
	CLAHE	**7.864**	0.650	3.045
Scence 1	MSRCR	6.507	0.346	2.999
	DCP	7.549	0.6262	**2.860**
	MSR-PO	7.835	**0.654**	3.013
	None	7.499	0.529	4.445
	CLAHE	**7.875**	0.592	**4.105**
Scence 2	MSRCR	5.899	0.309	4.987
	DCP	7.5643	0.595	4.253
	MSR-PO	7.840	**0.628**	4.340
	None	6.608	0.343	11.365
	CLAHE	7.329	0.438	9.660
Scence 3	MSRCR	6.303	0.434	8.178
	DCP	7.019	0.382	10.560
	MSR-PO	**7.642**	**0.641**	**7.341**
	None	6.595	0.340	12.426
	CLAHE	7.217	0.406	11.149
Scence 4	MSRCR	6.413	0.421	9.731
	DCP	7.030	0.388	11.780
	MSR-PO	**7.731**	**0.642**	**8.489**

Table 2. Normalized indicator results.

Scene	Enhanced Algorithm	Entropy	UCIQE	NIQE	Average Quality Index
Scence 1	None	0.203	0.214	0.205	0.207
	CLAHE	**0.211**	0.224	0.195	0.210
	MSRCR	0.174	0.120	0.197	0.164
	DCP	0.202	0.216	**0.207**	0.208
	MSR-PO	0.210	**0.226**	0.196	**0.211**
Scence 2	None	0.204	0.200	0.198	0.201
	CLAHE	**0.215**	0.223	**0.215**	**0.218**
	MSRCR	0.161	0.116	0.177	0.151
	DCP	0.206	0.224	0.207	0.212
	MSR-PO	0.214	**0.237**	0.203	**0.218**
Scence 3	None	0.189	0.153	0.162	0.168
	CLAHE	0.210	0.195	0.190	0.198
	MSRCR	0.181	0.194	0.224	0.200
	DCP	0.201	0.171	0.174	0.182
	MSR-PO	**0.219**	**0.287**	**0.250**	**0.252**
Scence 4	None	0.189	0.155	0.169	0.171
	CLAHE	0.206	0.185	0.189	0.193
	MSRCR	0.183	0.191	0.216	0.197
	DCP	0.201	0.177	0.178	0.185
	MSR-PO	**0.221**	**0.292**	**0.248**	**0.253**

7. Conclusions

The existing MSR-based underwater image enhancement algorithms use fixed parameters, and these algorithms are unstable when facing different underwater environments. Therefore, it is important to determine the appropriate parameters according to the changes of different scenarios.

MSR-PO chooses the appropriate NR-IQA method and optimizes the parameters using the MSR algorithm to achieve stable underwater image enhancement, making further underwater exploration possible. Secondly, we also consider the engineering application performance of this method, and we propose a way to restore the general parameters of short-time stable frames.

MSR-PO can complete the optimization and restoration of 30 frames within 1 s. At the same time, we reserve enough computing resources for the AUV's own analysis and control. This indicates that the MSR-PO algorithm can be applied in engineering applications. By comparing different scenes, the experimental results show that MSR-PO algorithm's subjective effect and objective quality scores are better than the existing classical MSR based algorithm and DCP algorithm. This shows that MSR-PO can be applied to different underwater environments and the effect is stable.

Author Contributions: Conceptualization, K.H., Y.Z., F.L., Z.D. and Y.L.; methodology, K.H., Y.Z. and F.L.; software, Y.Z. and F.L.; validation, K.H., Y.Z. and F.L.; formal analysis, K.H., Y.Z. and F.L.; investigation, K.H., Y.Z. and F.L.; resources, K.H., Y.Z. and F.L.; data curation, K.H., Y.Z. and F.L.; writing—original draft preparation, K.H., Y.Z. and F.L.; writing—review and editing, K.H., Y.Z. and F.L.; visualization, K.H., Y.Z. and F.L.; supervision, K.H., Y.Z. and F.L.; project administration, K.H.; funding acquisition, K.H. All authors have read and agreed to the published version of the manuscript.

Funding: The code used to support the findings of this study are available from the corresponding author upon request (nuistpanda@163.com). The data are from the open data set of the National Natural Science Foundation of China Underwater Robot Competition. (http://www.cnurpc.org/index.html).

Acknowledgments: The research in this paper is supported by the National Natural Science Foundation of China (61773219, 61701244), and the key special project of the National Key R&D Program (2018YFC1405703). Heartfelt thanks are expressed to the reviewers who submitted valuable revisions to this article.

Conflicts of Interest: No potential conflict of interest was reported by the author.

Abbreviations

The following abbreviations are used in this manuscript:

OMS-PO	MSR Parameter Optimization
MSR	Multi-Scale Retinex
NR-IQA	Non-Reference Image Quality Assessment
NIQE	Natural Image Quality Evaluator
GSA	Gravitational Search Algorithm
HE	Histogram Equalization
AHE	Adaptive Histogram Equalization
CLAHE	Contrast Limit Adaptive Histogram Equalization
MSRCR	Multi-Scale Retinal with Color Recovery
MSR	Multi-Scale Retinex
DCP	Dark Channel Prior
AUV	Autonomous Underwater Vehicle
UVMS	Underwater Vehicle Manipulator System
GAN	Generative Adversarial Networks
SSR	Single-Scale Retinex
MOS	Mean Opinion Score
DMOS	Differential Mean Opinion Score
FR-IQA	Full Reference Image Quality Assessment
RR-IQA	Reduced Reference Image Quality Assessment
UCIQE	Underwater Color Image Quality Evaluation
NIQE	The Natural Image Quality Evaluator
UIQM	Underwater Image Quality Measurement
UICM	Underwater Image Colorfulness Measurement
UISM	Underwater Image Sharpness Measurement
UIConM	Underwater Image Contrast Measurement
NFC	National Fund of China

References

1. Lin, X.; Wu, J.; Qin, Q. A Novel Obstacle Localization Method for an Underwater Robot Based on the Flow Field. *J. Mar. Sci. Eng.* **2019**, *7*, 437.
2. Chiarella, D.; Bibuli, M.; Bruzzone, G.; Caccia, M.; Ranieri, A.; Zereik, E.; Marconi, L.; Cutugno, P. A novel gesture-based language for underwater human–robot interaction. *J. Mar. Sci. Eng.* **2018**, *6*, 91.
3. Xia, M.; Song, W.; Sun, X.; Liu, J.; Ye, T.; Xu, Y. Weighted Densely Connected Convolutional Networks for Reinforcement Learning. *Intern. J. Pattern Recognit. Artif. Intell.* **2020**, *34*, 2052001.
4. Xia, M.; Wang, K.; Zhang, X.; Xu, Y. Non-intrusive load disaggregation based on deep dilated residual network. *Electr. Power Syst. Res.* **2019**, *170*, 277–285.
5. Xia, M.; Qian, J.; Zhang, X.; Liu, J.; Xu, Y. River segmentation based on separable attention residual network. *J. Appl. Remote Sens.* **2019**, *14*, 032602.
6. Xia, M.; Zhang, X.; Weng, L.; Xu, Y. Multi-stage Feature Constraints Learning for Age Estimation. *IEEE Trans. Inf. Forensics Secur.* **2020**, *15*, 2417–2428.
7. Xia, M.; Wang, K.; Song, W.; Chen, C.; Li, Y. Non-intrusive load disaggregation based on composite deep long short-term memory network. *Expert Syst. Appl.* **2020**, *160*, 113669.
8. Singhai, J.; Rawat, P. Image enhancement method for underwater, ground and satellite images using brightness preserving histogram equalization with maximum entropy. In Proceedings of the International Conference on Computational Intelligence and Multimedia Applications, Sivakasi, India, 13–15 December 2007.
9. Pizer, S.; Amburn, E.; Austin, J.; Cromartie, R.; Geselowitz, A.; Greer, T.; ter Haar Romeny, B.; Zimmerman, J.; Zuiderveld, K. Adaptive histogram equalization and its variations. *CVGIP* **1987**, *39*, 355–368.
10. Zuiderveld, K. Contrast limited adaptive histogram equalization. In *Graphics Gems IV*; Academic Press Professional, Inc.: San Diego, CA, USA, 1994; ISBN 012-336-155-9.

11. Liu, Y.; Chan, W.; Chen, Y. Automatic white balance for digital still camera. *IEEE Trans. Consum. Electron.* **1995**, *41*, 460–466.

12. Van De Weijer, J.; Gevers, T.; Gijsenij, A. Edge-based color constancy. *IEEE Trans. Image Process.* **2007**, *16*, 2207–2214.

13. Land, E.; McCann, J. Lightness and retinex theory. *J. Opt. Soc. Am.* **1971**, *61*, 1–11.

14. Zhang, S.; Wang, T.; Dong, J.; Yu, H. Underwater image enhancement via extended multi-scale Retinex. *Neurocomputing* **2017**, *245*, 1–9.

15. Jobson, D.; Rahman, Z.; Woodell, G. A multiscale retinex for bridging the gap between color images and the human observation of scenes. *IEEE Trans. Image Process.* **1997**, *6*, 965–976.

16. Rahman, Z.; Jobson, D.; Woodell, G. Multiscale retinex for color rendition and dynamic range compression. *Appl. Digit. Image Process. XIX* **1996**, *2847*, 183–191.

17. Tang, C.; von Lukas, U.; Vahl, M.; Wang, S.; Wang, Y.; Tan, M. Efficient underwater image and video enhancement based on Retinex. *Signal Image Video Process.* **2019**, *13*, 1011–1018.

18. Zhang, W.; Li, G.; Ying, Z. A new underwater image enhancing method via color correction and illumination adjustment. In Proceedings of the 2017 IEEE Visual Communications and Image Processing(WCIP), St. Petersburg, FL, USA, 10–13 December 2017.

19. Mercado, M.; Ishii, K.; Ahn, J. Deep-sea image enhancement using multi-scale retinex with reverse color loss for autonomous underwater vehicles. In Proceedings of the OCEANS 2017-Anchorage, Anchorage, AK, USA, 18–21 September 2017.

20. Dai, C.; Lin, M.; Wang, J.; Hu, X. Dual-purpose method for underwater and low-light image enhancement via image layer separation. *IEEE Access* **2019**, *7*, 178685–178698.

21. Xiao, C.; Gan, J. Fast image dehazing using guided joint bilateral filter. *Vis. Comput.* **2012**, *28*, 713–721.

22. Li, C.; Guo, J.; Cong, R.; Pang, Y.; Wang, B. Underwater image enhancement by dehazing with minimum information loss and histogram distribution prior. *IEEE Trans. Image Process.* **2016**, *25*, 5664–5677.

23. Chiang, J.; Chen, Y. Underwater image enhancement by wavelength compensation and dehazing. *IEEE Trans. Image Process.* **2011**, *21*, 1756–1769.

24. Zhu, J.; Park, T.; Isola, P.; Efros, A. Unpaired image-to-image translation using cycle-consistent adversarial networks. In Proceedings of the IEEE International Conference on Computer Vision, Venice, Italy, 22–29 October 2017.

25. Huang, Z.; Wan, L.; Sheng, M.; Zou, J.; Song, J. An Underwater Image Enhancement Method for Simultaneous Localization and Mapping of Autonomous Underwater Vehicle. In Proceedings of the 2019 3rd International Conference on Robotics and Automation Sciences (ICRAS), Wuhan, China, 1–3 June 2019.

26. Chen, W.; Wang, L.; Zhang, Y.; Li, X.; Liu, J.; Wang, W. Anti-disturbance grabbing of underwater robot based on retinex image enhancement. In Proceedings of the 2019 Chinese Automation Congress (CAC), Hangzhou, China, 22–24 Novembe 2019.

27. Rashedi, E.; Nezamabadi-Pour, H.; Saryazdi, S. GSA: A gravitational search algorithm. *Inf. Sci.* **2009**, *179*, 2232–2248.

28. Jobson, D.; Rahman, Z.; Woodell, G. Properties and performance of a center/surround retinex. *IEEE Trans. Image Process.* **1997**, *6*, 451–462.

29. Liu, T.; Lin, Y.; Lin, W.; Kuo, C. Visual quality assessment: Recent developments, coding applications and future trends. *APSIPA Trans. Signal Inf. Process.* **2013**, *2*, e4.

30. Liu, X.; van de Weijer, J.; Bagdanov, A. Rankiqa: Learning from rankings for no-reference image quality assessment. In Proceedings of the IEEE International Conference on Computer Vision, Venice, Italy, 25 December 2017.

31. Shannon, C. A mathematical theory of communication. *Bell Syst. Tech. J.* **1948**, *27*, 379–423.

32. Yang, M.; Sowmya, A. An underwater color image quality evaluation metric. *IEEE Trans. Image Process.* **2015**, *24*, 6062–6071.

33. Mittal, A.; Soundararajan, R.; Bovik, A. Making a "completely blind" image quality analyzer. *IEEE Signal Process. Lett.* **2012**, *20*, 209–212.

34. Panetta, K.; Gao, C.; Agaian, S. Human-visual-system-inspired underwater image quality measures. *IEEE J. Ocean. Eng.* **2015**, *41*, 541–551.

Journal of
*Marine Science
and Engineering*

Article

On the Estimation of the Surface Elevation of Regular and Irregular Waves Using the Velocity Field of Bubbles

Diana Vargas [1], Ravindra Jayaratne [2], Edgar Mendoza [1,*] and Rodolfo Silva [1]

[1] Instituto de Ingeniería, Universidad Nacional Autónoma de México, Cd. Universitaria, CdMx 04510, Mexico; karolz_27@yahoo.com.mx (D.V.); rsilvac@iingen.unam.mx (R.S.)

[2] School of Architecture, Computing & Engineering, University of East London, Docklands Campus, 4–6 University Way, London E16 2RD, UK; r.jayaratne@uel.ac.uk

* Correspondence: emendozab@iingen.unam.mx; Tel.: +52-5556-233600 (ext. 8957)

Received: 30 December 2019; Accepted: 29 January 2020; Published: 2 February 2020

Abstract: This paper describes a new set of experiments focused on estimating time series of the free surface elevation of water (FSEW) from velocities recorded by submerged air bubbles under regular and irregular waves using a low-cost non-intrusive technique. The main purpose is to compute wave heights and periods using time series of velocities recorded at any depth. The velocities were taken from the tracking of a bubble curtain with only one high-speed digital video camera and a bubble generator. These experiments eliminate the need of intrusive instruments while the methodology can also be applied if the free surface is not visible or even if only part of the depth can be recorded. The estimation of the FSEW was successful for regular waves and reasonably accurate for irregular waves. Moreover, the algorithm to reconstruct the FSEW showed better results for larger wave amplitudes.

Keywords: free surface recording; waves; bubbles; PIV; wave flume experiments

1. Introduction

The kinematic study of the free surface elevation of water (FSEW) is the main data source to estimate the wave velocities and forces acting on maritime infrastructure. The measurement of the FSEW with non-intrusive techniques is one of the most challenging topics in the field of Coastal Engineering and Fluid Dynamics. Over the last 30 years, a vast number of research studies on the behavior of the FSEW of irregular waves have been carried out. The authors of [1] presented a comparison between mathematical and laboratory measurements of velocities below the FSEW with a laser Doppler velocimeter (LDV) to compute the wave forces on submarine pipelines. In this work, water level fluctuations were measured to calculate the time series of velocities. The same set of experiments was replicated with inductive- and impeller-type probes instead of an LDV to confirm the theories with the prediction of velocities and pressures by [2,3]. These authors also estimated the FSEW using pressure data. In [4], a transfer function was obtained to estimate the surface from pressure measurements in deep water. Later, [5] compared the theory and measurements of water particle velocities in solitary waves, while [6] carried out the same experiments under monochromatic waves.

Calculations of spectral transfer functions between the FSEW and subsurface 3D particle velocities in wind-induced waves were also performed by [7]. Other researchers also measured water particle velocities and water surface elevations [8–11]. In addition, measurements of pressure under irregular waves were taken to estimate the water depth at several positions [12,13]. Pressure-, capacitance-, and resistance-type wave gauges were used by [14] in traditional detection of FSEW.

These techniques usually constitute discrete localized point measurements; hence, the recorded data are not enough to understand the behavior of the water surface. Due to these limitations, new optical techniques to measure the FSEW were developed to obtain more accurate datasets regarding time and space. The authors of [15] created a new technique combining colorimetry and digital image processing to obtain the 3D free surface elevation for a time-dependent flow. The authors of [16] developed a new technique to measure two components of the water surface gradient and, subsequently, [17] created an algorithm for the estimation of water surface elevation by integrating the surface gradient.

By that time, the measurements of velocity fields (e.g., [18]) were obtained using imaging techniques, such as laser speckle velocimetry (LSV), particle imaging velocimetry (PIV), particle tracking velocimetry (PTV), defocusing digital particle image velocity (DPIV), stereo photography, shadowgraphy, laser slope gauges, and optical displacement sensors as diffusing light photography by [19,20], who measured deflection of a reflected laser beam on the free surface at each point.

Fringe projection profilometry was used by [21] to measure the FSEW, while [22] broadened the study into the well-known Fourier transform profilometry (FTP). This technique was later combined with digital PIV [23,24] to study the relationship between small-sloped FSEW deformations and near-surface velocities. Based on laser speckle techniques, [25] developed a system to measure the distortion of a collimated speckle pattern. In addition to these techniques, [26,27] developed two different stereoscopic methods for measuring 2D water surface elevations. [28] used dynamic refraction stereo to estimate an unknown refractive 3D water surface. More recently, [29] used phase shifting and FTP to estimate the FSEW in a 3D flow. Later, [30,31] performed simultaneous measurements of topography and velocity fields on FSEW and [32] analyzed a refracted image and a reference image to measure the FSEW. Other research studies measured the deformation of the FSEW with a modification to the optical profilometry technique [33] and measured the flow fields beneath the free surface of the waves with PTV [34,35]; [36,37] also used PTV to measure wave-induced mean flow and particle trajectories for linear wave packets. In turn, [38] used PIV methods to evaluate mass transport velocity based on measured vector fields. A theoretical study recently was made to predict the surface waves following the recovery of the pressure at the bottom [39].

It can thus be seen a growing interest in understanding and characterizing the FSEW to find the relationships between wave hydrodynamics and other related parameters such as wave energy, pressure, velocities, and vertical acceleration below the free surface, the forces which deform the FSEW in the presence of coastal structures, etc.

Despite these efforts to improve the measurements, some methods interfere with the FSEW. For example, wave gauges can act as obstacles for the free surface, consequently altering the behavior of the wave. On the other hand, non-intrusive water surface visualization has been limited to optical techniques [40]. Bubble image velocimetry (BIV) techniques have been adopted by various researchers to obtain the velocity of bubbles generated by air entrapment and entrainment of breaking and broken waves to study wave hydrodynamics and overtopping [41,42] as well as kinematics around pneumatic breakwaters [43]. Nevertheless, as the bubbles were of different sizes, shapes, and buoyancy forces, the application of the BIV technique or algorithm was hindered in some cases. Other research studies of the velocity of bubbles were carried out using electrolysis, the measurement of the vertical distributions of the water particle velocity induced at a certain wave phase [44]. Another measurement technique was using the defocusing of digital particle images to calculate the velocity field from the cross-correlation of volumes from two sequential defocusing bubble images [45]. There have also been studies on the characteristics of bubbles rising in water; i.e., velocity, trajectory, oscillation frequency, size, and density [46–49].

The main purpose of the present study is to use a non-intrusive technique that measures the velocities of a curtain of artificially created bubbles. The bubbles were generated with compressed air using a device made in-house that was placed at the bottom of the flume. The velocity of the bubbles is considered as a proxy of the water velocity and thus a transfer function was used to compute the

FSEW from the time series of the velocity recorded at a known depth. Conventional techniques (wave gauges, UVP, and PIV) were used to validate the proposed methodology and to establish the depth and period ranges for which better results were obtained.

2. Materials and Methods

The experimental program was carried out in the wave flume of the National Autonomous University of Mexico (UNAM), which is 37.0 m long, 0.79 m wide, and 1.2 m deep. The flume is constructed of tempered glass walls and the wave paddle is equipped with an active absorption system to reduce the re-reflected waves. The bubble device consisted of an acrylic tube 0.80 m long, a diameter of 0.0254 m and 0.006 m thick. The device had 1 mm diameter holes separated half a centimeter along its length. The ends of the tube were hermetically sealed, and the air was supplied by a compressor connected to the central part of the tube. The pressure of the compressor was set to ensure the proper number, size, and visibility of the bubbles.

The bubble generator was placed at the bottom of the flume, centered along the channel and parallel to the wave direction. The distance from the device to the dissipative beach at the end of the flume, as shown in Figures 1 and 2, was 9 m. To record the motion of the bubbles, a high-speed digital video camera set to 12,000 fps and a lens 25 mm wide were located parallel to the flume, 3.7 m away from the curtain of bubbles. The bubble curtain was illuminated by LED lamps and a black background was placed to enhance the contrast. Five wave gauges were located along the flume. The first one was placed 4.0 m away from the wavemaker paddle as a witness of the generated waves and the other four gauges were placed within the bubble curtain.

Three velocity profiles were recorded using an ultrasonic velocity profiler (UVP) located close to each of the four-wave gauges and laser PIV velocity maps were used to validate the velocity fields obtained tracking the air bubbles.

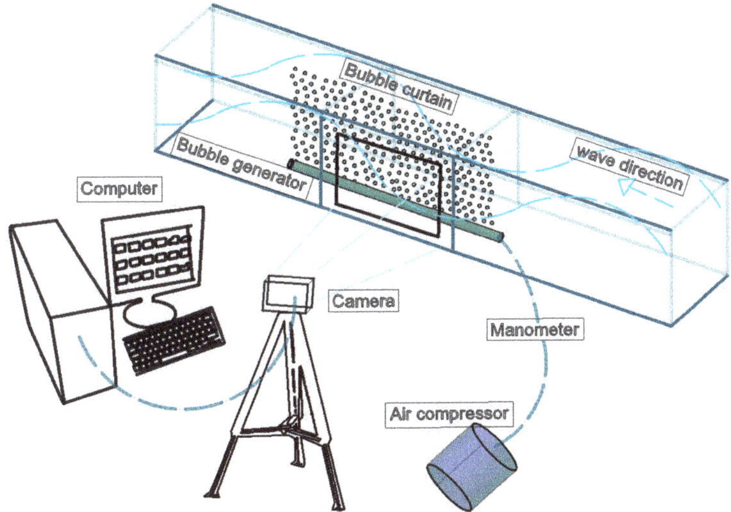

Figure 1. Experimental setup used to record the velocity of air bubbles beneath superficial waves.

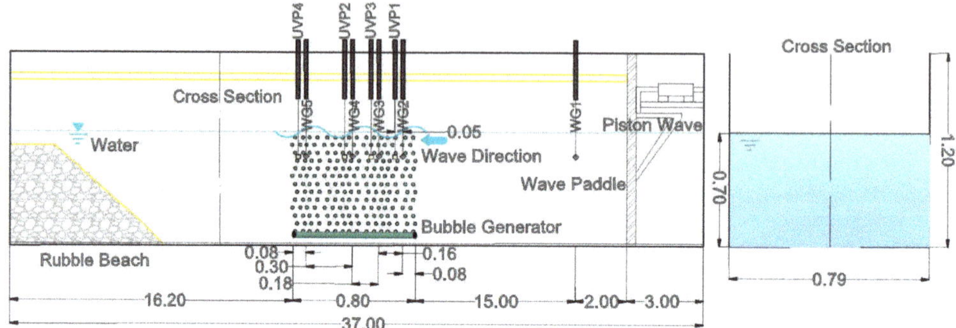

Figure 2. Wave flume cross section. WG stands for wave gauge and UVP for ultrasonic velocity profiler. Dimensions are in meters.

3. Methodology

The experiments were performed with a still water reference depth of 0.7 m. The FSEW was perturbed by regular and irregular waves (Jonswap spectrum with $\gamma = 3.3$) with different wave heights and periods as given in Table 1. The wave trains were chosen focusing on covering a wide range of conditions, always within linear wave conditions, in order to achieve sensibility regarding the applicability of the technique for estimation of FSEW. A total of 32 tests were conducted, 16 different waves trains (8 regular and 8 irregular), were measured and each test was repeated two times.

Table 1. Wave conditions used in present study. R and I denote regular and irregular waves, respectively.

Case No	Peak Period T_p (s)	Significant Wave Height H_s (m)
R1/I1	0.8	0.01
R2/I2	0.8	0.03
R3/I3	0.8	0.05
R4/I4	0.8	0.10
R5/I5	1.2	0.01
R6/I6	1.2	0.03
R7/I7	1.2	0.05
R8/I8	1.2	0.10

To record the motion of the bubbles, the high-speed digital video camera was focused on the plane of the bubble curtain and series of 10,390 images were recorded. The pre-processing of the imagery consisted on three stages to enhance the contrast and highlight and define the bubbles, particularly the smallest bubbles that seemed blurred. First, the size of the picture was trimmed from 644 × 400 pixels (original size of the picture) to 397 × 222 pixels to capture the study area. Second, the background was captured without bubbles, and its average was subtracted from the images to eliminate any other source of light that does not correspond to the bubbles. Third, the bubbles were recorded without waves to calculate an average velocity field of the ascending motion that was subtracted from the velocity fields of the bubbles with waves.

To scale the images, a metric scale was located in the plane of the bubble curtain to find the conversion factor to be introduced later in the calculations of the velocity field. The size of a pixel in the picture resulted in 1 px = 0.00195 ± 0.0005 m. Given that the sampling frequency was 250 Hz, the conversion factor px/frame to m/s was estimated as 0.4875 ± 0.0005 m/s and the duration of each test was 41.56 s. The duration of the test allowed to record 34 waves of 1.2 s. The wave gauges were set with a sampling frequency of 100 Hz (accuracy of the sensors is 0.1 mm).

These photographs were analyzed with PIVLab code [50], employing a single frame and three interrogation window sizes: 64 × 32, 32 × 16, and 16 × 8 pixels to construct the velocity field. Additionally, the correlations between windows were made with a Fourier transform, a deformation linear window, and a Gaussian subpixel estimator. For the missing vectors, a spline interpolation method was applied. The total number of photographs in each test was 10390; this means that after applying the PIVLab analysis, 5195 velocity fields were obtained which are more than enough for the proposed analysis. The time series values were obtained as the average of a selected region at a fixed depth. Figure 3 shows an image of a velocity field where different depths have been marked; time series of velocities were taken at these depths.

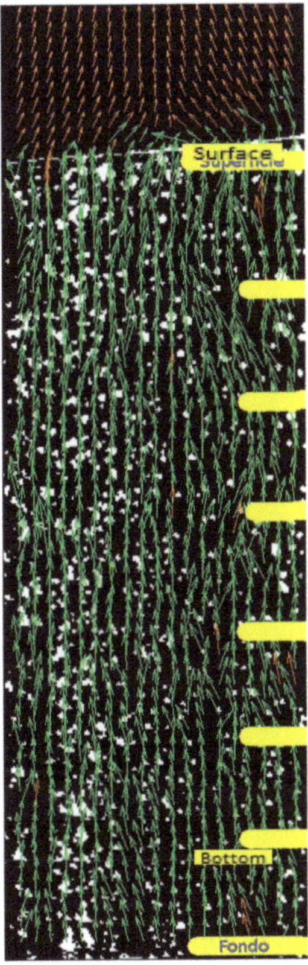

Figure 3. Photograph processed with particle imaging velocimetry (PIV). The green vectors are the velocities of the bubbles and the red vector are the vectors interpolated due to a missing data.

For validation purposes, the FSEW was recorded both with the wave gauges and from the photographs using the Sobel operator to detect edges or borders in images.

Singular spectrum analysis (SSA) was used to remove high-frequency oscillations from the time series, and its Fourier spectrum was calculated for the two time series: one from the velocity of the bubbles (sampled at 250 Hz) and other from the wave gauges (sampled at 100 Hz). As the sampling frequencies were dissimilar, one point every 5 and one point every 2 were used from the bubble and wave gauge time series, respectively, for the comparison. A linear adjustment between the coefficients of the harmonic components of the Fourier spectra of the two time series was computed. This adjusted relationship corresponds to the transfer function which allows the FSEW to be estimated from the time series of the velocities of the bubbles.

4. Results and Discussion

To obtain the pixels corresponding to the position of the FSEW, the contour found for the still water level was used as a reference. Then, the FSEW recorded by the wave gauges and the contour in the photographs were compared.

Figure 4 compares the two time series (wave gauges and contours in the recorded images) of the FSEW. The mean absolute percentage error resulted in 5.5%, showing that an accurate estimation of the FSEW from the photographs is easy to achieve. Between the records, a phase shift exists due to the different start of the records and, therefore, an algorithm was created to match the initial time of each record and accomplish an appropriate comparison.

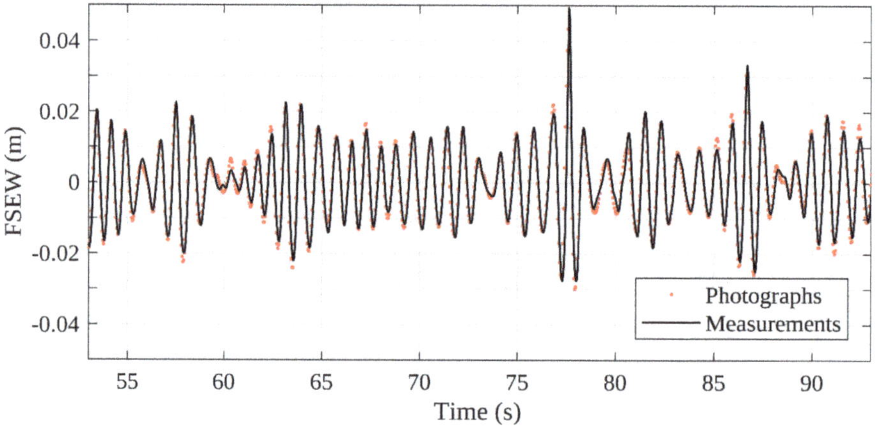

Figure 4. Comparison of free surface elevation of water (FSEW) recorded by wave gauges and extracted from the images for $T_p = 0.08$ s and $H_s = 0.05$ m, case I3-05.

The main results were derived from the time series of velocity field from the PIVLab analysis. Figure 5 presents the time series of the regular waves case $T = 0.8$ s and $H = 0.05$ m, whereas Figure 6 shows the data for irregular waves with $T_p = 0.8$ s and $H_s = 0.05$ m. It is important to mention that these time series were constructed using all the pixels across the width of the photograph (222 pixels or 0.433 m) at 4 cm depth. The values estimated for H and T from the FSEW obtained via the image analysis were very similar to the actual generated values. The error was found to be inversely proportional to wave height and period.

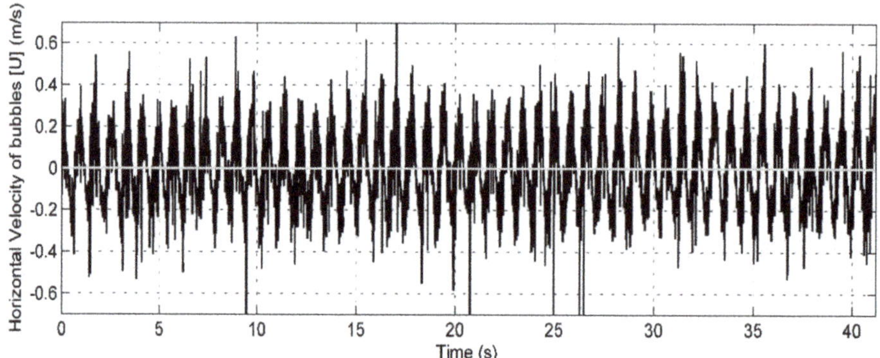

Figure 5. Time series of the horizontal velocity of the bubbles for regular waves with $H = 0.05$ m and $T = 0.8$ s; the depth is 4 cm.

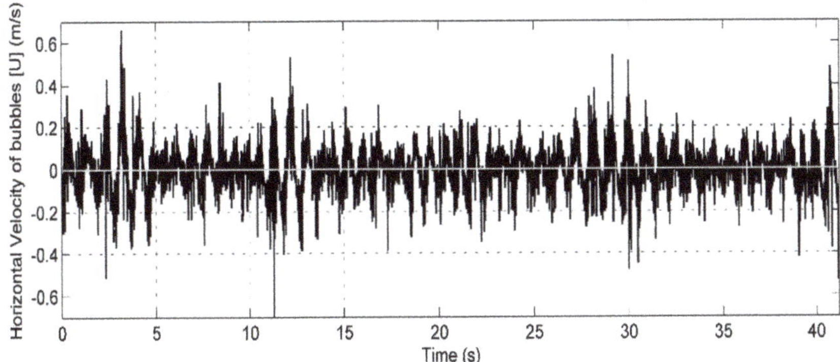

Figure 6. Time series of the horizontal velocity of the bubbles for irregular wave with H_s = 0.05 m and T_p = 0.8 s; the depth is 4 cm.

To understand the scope of the technique, the wave heights and the periods were classified into three types according to the amplitude and period as high (0.10 m, 1.2 s), medium (0.05 m, 0.8 s), and small (0.01 m, 0.8 s), as illustrated in Table 2.

Table 2. Summary of experimental wave parameters used in present study.

Case No	Peak Period T_p (s)	Significant Wave Height H_s (m)	Ursell Number	H/gT^2	h/gT^2	Wave Theory
R3/I3	0.8	0.05	0.145	0.008	0.111	Stokes 2nd
R4/I4	0.8	0.10	0.290	0.016	0.111	Stokes 3rd
R7/I7	1.2	0.05	0.687	0.003	0.050	Stokes 2nd
R8/I8	1.2	0.10	1.374	0.007	0.050	Stokes 3rd

The Ursell number and two non-dimensional parameters (H/gT^2 and h/gT^2) were calculated to identify the Stokes order of each case. This was carried out regardless of the behavior of the waves and, hence, the bubbles detected the presence of waves.

The Fourier spectra of the time series of the bubbles and the FSEW obtained from the contours of the pictures, for regular waves, were computed to compare the frequencies in one spectrum and the other. For that reason, 20 spectra at different water depths were calculated and normalized by the maximum value.

In Figure 7a,b, the bubbles in regular waves responded according to the perturbation caused due to waves because the 20 spectra had the same shape among them, the same bandwidth, and their difference in magnitude was not large. This means that the time series of the velocities of the bubbles immediately below the FSEW gave better results than the time series close to the bottom. Figure 7e shows the Fourier spectrum obtained by the time series measured by the photographs (or if it is the case by the wave gauges) which agrees with the Fourier spectrum obtained by the 20 times series of the velocities of the bubbles (Figure 7a,b).

Figure 7c,d,f shows the Fourier spectra for irregular wave test cases. In this case, the spectra show multiple peaks. It is important to note that the maximum peak frequency is in good agreement with the initial frequency of the experiment. As mentioned before, 20 Fourier spectra were calculated corresponding to different water depths. The behavior of the frequencies shows that the energy decreases as the water depth decreases, but the peak frequency is always present in each spectrum meaning that up to a certain water depth, the bubbles result in good detection of the peak period of the waves.

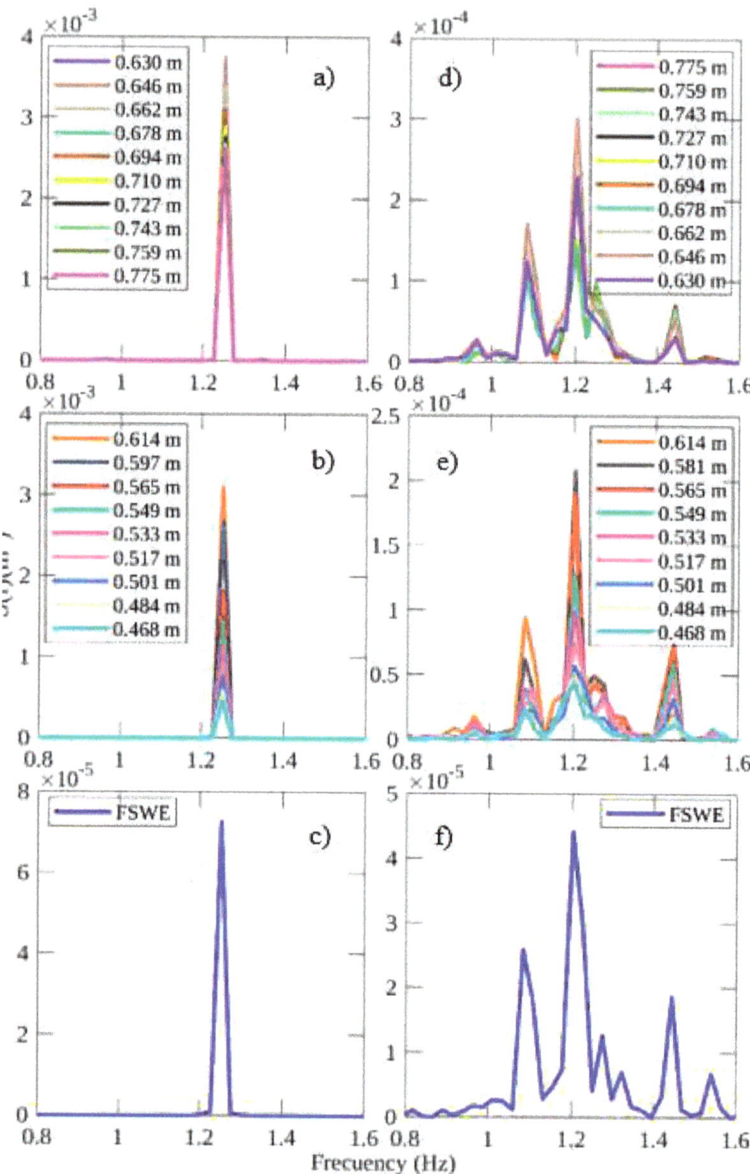

Figure 7. Fourier spectra for (**a**) regular waves $H = 0.05$ m and $T = 0.08$ s using the time series of the velocities of the bubbles for depths ranging from 0.630 to 0.775; (**b**) regular waves $H = 0.05$ m and $T = 0.08$ s using the time series of the velocities of the bubbles for depths ranging from 0.468 to 0.614 m; (**c**) regular waves $H = 0.05$ m and $T = 0.08$ s using the time series of the FSEW obtained from the contours of the photographs; (**d**) irregular waves $H_s = 0.05$ m and $T_p = 0.08$ s using the time series of the velocities of the bubbles for depths ranging from 0.630 to 0.775 m; (**e**) irregular waves $H_s = 0.05$ m and $T_p = 0.08$ s using the time series of the velocities of the bubbles for depths ranging from 0.468 to 0.614 m; and (**f**) irregular waves $H_s = 0.05$ m and $T_p = 0.08$ s using the time series of the FSEW obtained from the contours of the photographs.

After the calculation of the FFT of the time series of the velocities of the bubbles and of the FFT of the FSEW measured from the contours of the photographs, a scatter plot was made for the set of data to fit a transfer function, which allows computing the estimation of the FSEW. This process was repeated for regular (Figure 8) and irregular waves (Figure 9). In addition, the squared linear correlation coefficient (R^2) was obtained for different water depths, going from the surface (4 cm depth) to the bottom (0.347 m depth). From these, it can be observed that the technique works better for the series of velocities of the bubbles which are found closer to FSEW. This is because the forcing of the wave diminishes with the depth. Hence, bubbles at the bottom are less influenced by the oscillatory motion than the bubbles in smaller depths.

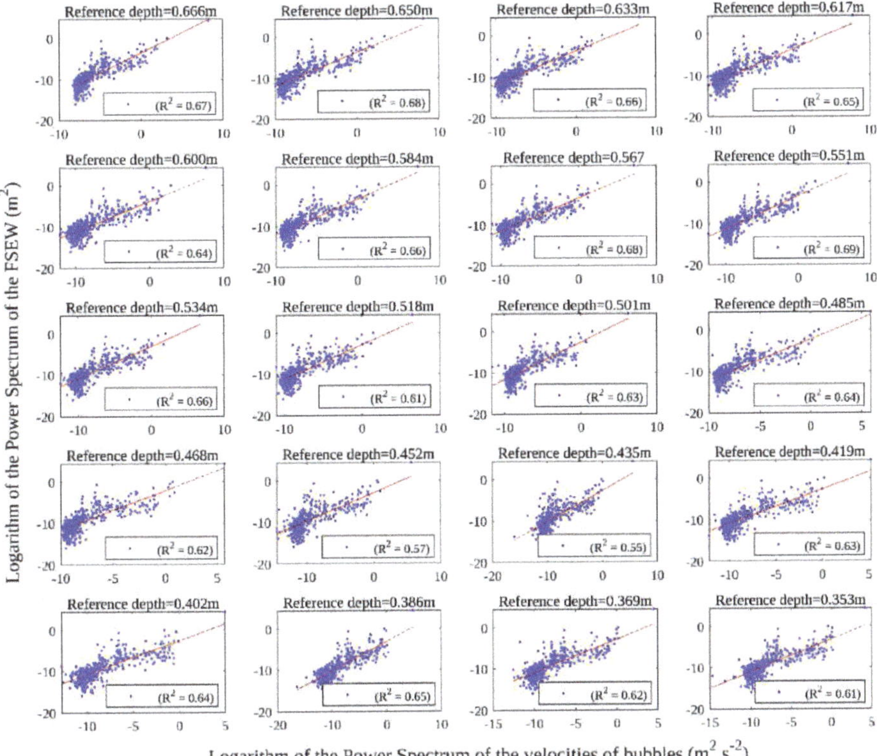

Figure 8. Relationship between the Fourier spectrum of the time series of the velocities of the bubbles and the Fourier spectrum of the FSEW measured for regular waves with $H = 0.05$ m and $T = 0.08$ s.

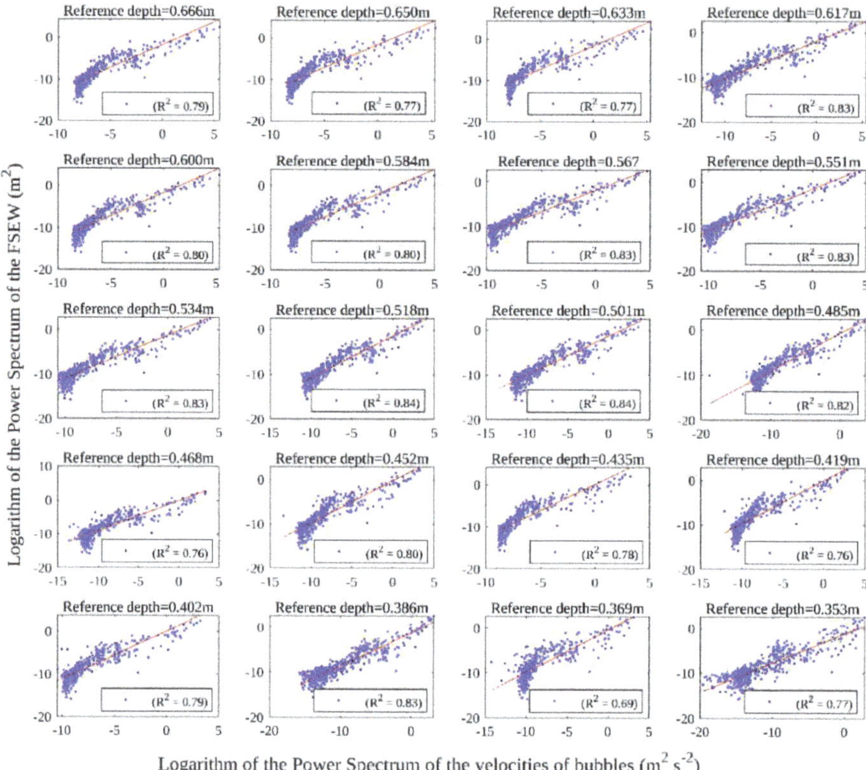

Figure 9. Relationship between the Fourier spectrum of the time series of the velocities of the bubbles and the Fourier spectrum of the FSEW measured for irregular waves with $H_s = 0.05$ m and $T_p = 0.08$ s.

The range of valid time series along the depth was chosen from 0.04 to 0.347 m depth. Closer to the free surface than the first limit, the PIV velocity fields are not properly estimated due to the lack of bubble neighbors in the images. The second limit is the threshold for which acceptable error was found in the wave period results. The R^2 values in Figures 8 and 9 are similar for the different water depths and no trend can be seen, indicating that R^2 is independent of the depth; this means that the technique can be used with the same accuracy for any depth within the valid range. Globally, despite R^2 having consistent values, it is also observed that especially for irregular waves, some frequencies are far from the linear adjustment and it has not been studied yet if any of these frequencies can improve the estimation of the FSEW.

The estimations of the FSEW were obtained from velocity data at different depths with transfer functions. These transfer functions were calculated by adjusting the power spectra of the time series of the velocities of the bubbles and the power spectra of the time series of the FSEW obtained with the contours from the photographs. The linear adjustment between the two power spectra in a log–log scale relates the coefficients of the amplitude of each Fourier harmonic between the two spectra. Hence, each harmonic of the time series of the velocity of the bubbles was adjusted to obtain the values of each harmonic of the FSEW. The power spectrum is defined as the absolute value of the square of the Fourier transform coefficients.

As shown in Figures 10 and 11, for regular waves, the estimated (Reconstructed) and measured FSEW obtained from the contours of the photographs (Target) are similar, except for 0.29 m depth. Hence, it is found that this depth is a threshold depth that defines where the technique stops working.

For the irregular waves, the most significant difference between the reconstructed and target curves occurred when the FSEW reaches its largest absolute values, or when there is a small or large change in the amplitude. The comparison of the FSEW can be considered accurate enough because the mean error calculated with the six-time series presented in Figure 10 was of ±0.0025 m for regular waves and ±0.004 m for irregular waves. The mean error is defined as the mean of the absolute value of the subtraction between the two curves (FSEW Target − FSEW Reconstructed). If it is considered H = 0.05 m, then the error for regular waves was 5% and for irregular waves was 8%.

Figure 12 summarizes the results obtained for the cases of the Table 1 for irregular waves. There, the relationship between R^2 and the Relative Depth (0 corresponds to the still water level and 1 corresponds to the bottom) is shown. The term R^2 is defined as the squared linear correlation coefficient of the reconstructed FSEW (Reconstructed) and the measured FSEW (Target). For the cases where the wave heights are 0.1 and 0.05 m and the periods are 0.8 and 1.2 s, a clear trend is observed such that as the depths decrease, the R^2 values decrease dramatically. The same abrupt change is found for wave heights of 0.03 and 0.01 m and periods of 0.8 and 1.2 s. Nevertheless, the curves for 0.01 m wave heights do not clearly show this abrupt change. However, for some water depths, the velocity field of the air bubbles is a good predictor of the FSEW.

The correlation values decrease with depth. Arguably, this is due to the closeness to the bubble generator where the velocity of the bubbles is dominated by their vertical displacement and thus no velocity related to the waves can be measured. The curves of Figure 12 reveal zones where the technique works, and the bubbles are sufficiently influenced by the waves such that they carry enough information to reconstruct the FSEW.

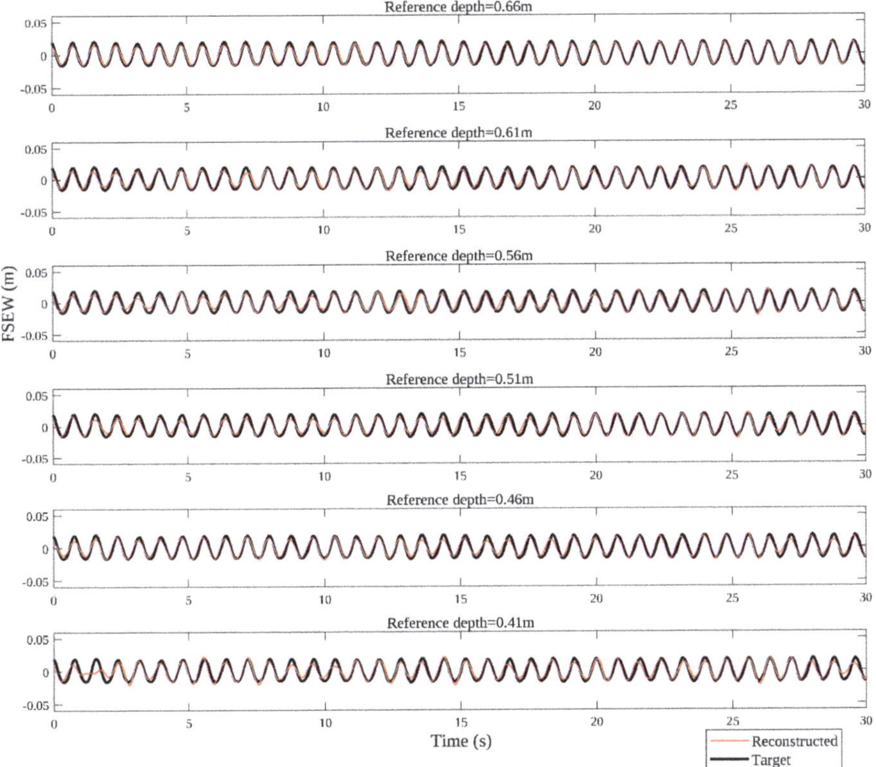

Figure 10. Measured and estimated FSEW for regular waves with H = 0.05 m and T = 0.08 s.

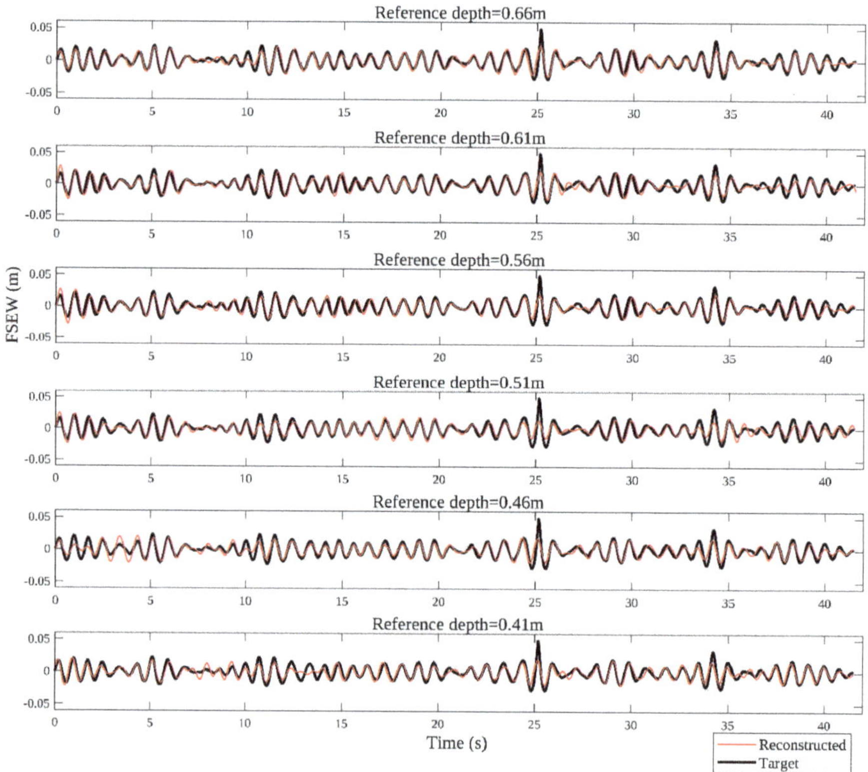

Figure 11. Measured and estimated FSEW for irregular waves with $H_s = 0.05$ m and $T_p = 0.08$ s.

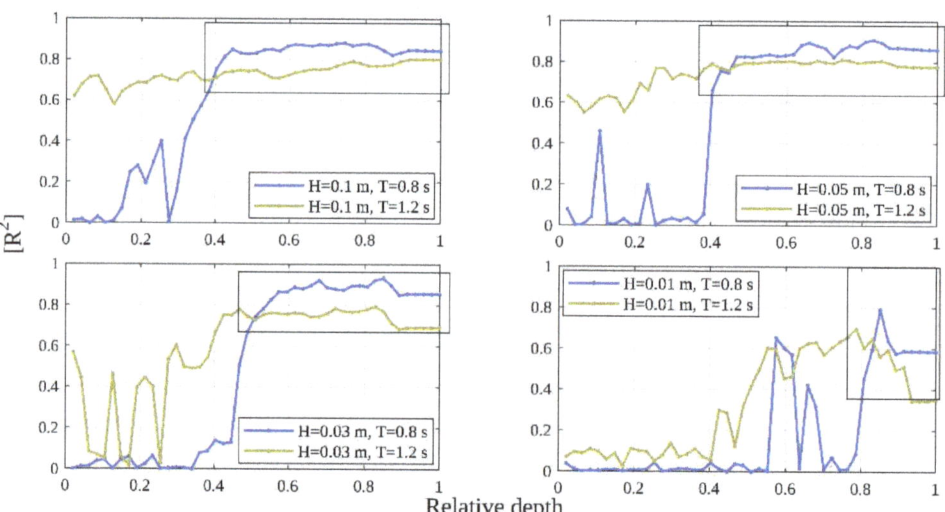

Figure 12. Critical curves defined as the squared linear correlation coefficients (R^2) as a function of relative depth for different wave periods and wave heights.

5. Conclusions

The technique employed to estimate the FSEW through the velocity field of the air bubbles gave good results (for a range of depths) compared to traditional measuring methods. The horizontal motion of the bubbles can also serve as a proxy to the motion of the water particles in coastal waves. Beyond reconstruction of the FSEW, the paper focuses on the relationship between the spectral peaks of the horizontal velocities of the air bubbles and of the FSEW.

Hence, it was observed that for regular waves, the oscillating movement of the air bubbles was in accordance with the wave period, whereas the velocities and periods of oscillation of the bubbles under irregular waves coincided with the range of the periods found in the irregular wave train. The critical curves obtained for different initial conditions describe the depths where the technique is valid. In our experiments, it was observed that for 0.7 m depth, wave amplitudes of 0.1 m, and periods of 1.2 s, the technique works well for depths lower than 0.4 m excluding subsurface area. It should be noted that for depths of around 0.03 m, the time series of the bubble velocities shows some irregularities. Arguably, this zone is behaving as a boundary layer between water and air.

The results show that the technique does not work very well to determine the FSEW when there is an abrupt change in the wave amplitude, for example, as shown in Figure 9 at 25 s.

The proposed non-intrusive methodology is useful in 2D wave flumes where it is desirable to obtain the FSEW without the use of typical wave gauges as well as to determine the behavior of the velocity of water particles that are below the FSEW. Moreover, the bubble generator is user-friendly and economical compared to a wave gauge. By contrast, it requires precise coupling between the speed of the camera, the control of bubble velocities, and the magnitude of fluid velocity.

A plan for future study is to perform more experiments within the range under which the technique is valid, in terms of the total depth, amplitudes, and periods. Moreover, it is recommended that the response time of the bubbles under different wave forcing effects be studied, for example, during wave breaking, where mass transport could play a significant role and could be important for FSEW reconstruction using the proposed technique.

Author Contributions: Conceptualization, D.V. and R.S.; methodology, R.J. and D.V.; software, D.V. and E.M.; validation, D.V. and E.M.; formal analysis, E.M. and R.J.; investigation, D.V.; resources, R.S. and R.J.; data curation, D.V.; writing—original draft preparation, D.V. and E.M.; writing—review and editing, R.J., E.M. and R.S.; visualization, D.V.; supervision, R.S. All authors have read and agreed to the published version of the manuscript.

Funding: This research received no external funding.

Acknowledgments: The data analysis of the proposed work was carried out at the University of East London (UEL) while spending postgraduate student internship of the first author. Therefore, the authors are grateful for the computing facilities and laboratory space provided by the UEL. The authors would also like to thank CEMIE-Oceano (CONACYT grant no. 249795) for partially funding this research.

Conflicts of Interest: The authors declare no conflict of interest.

References

1. Vis, F.C. Orbital velocities in irregular waves. In Proceedings of the 17th Coastal Engineering Conference, Sydney, Australia, 23–28 March 1980; pp. 173–185.
2. Daemrich, K.-F.; Eggert, W.-D.; Kolhase, S. Investigations on irregular waves in hydraulic models. In Proceedings of the 17th Coastal Engineering Conference, Sidney, Australia, 23–28 March 1980; pp. 186–203.
3. Daemrich, K.F.; Eggert, W.-D.; Cordes, H. Investigations on orbital velocities and pressures in irregular waves. In Proceedings of the 18th Coastal Engineering Conference, Cape Town, South Africa, 14–19 November 1982; pp. 297–311.
4. Lee, D.-Y.; Wang, H. Measurement of surface waves from subsurface gauge. In Proceedings of the 19th International Conference on Coastal Engineering, Houston, TX, USA, 3–7 September 1984; pp. 271–286.
5. Lee, J.-J.; Skjelbreia, J.M.; Raichlen, F. Measurement of velocities in solitary waves. *J. Waterw. Port C. Div.* **1982**, *108*, 200–218.

6. Bullock, G.N. Short I. Water particle velocities in regular waves. *J. Waterw. Port C. Ocean Eng.* **1985**, *111*, 89–200.

7. Battjes, J.A.; Van Heteren, J. Verification of linear theory for particle velocities in wind waves based on field measurements. *Appl. Ocean Res.* **1984**, *6*, 187–196. [CrossRef]

8. Swan, C. Convection within an experimental wave flume. *J. Hydraul. Res.* **1980**, *28*, 273–282. [CrossRef]

9. Hansen, J.B. Periodic waves in the surf zone: Analysis of experimental data. *Coast. Eng.* **1990**, *14*, 19–41. [CrossRef]

10. Kim, C.H.; Randall, R.E.; Krafft, M.J.; Boo, S.Y. Experimental study of kinematics of large transient wave in 2D wave tank. In Proceedings of the 22nd Annual Offshore Technology Conference, Houston, TX, USA, 7–10 May 1990; pp. 195–202.

11. Zhang, J.; Randall, R.E.; Spell, C.A. On wave kinematics approximate methods. In Proceedings of the 23rd Annual Offshore Technology Conference, Houston, TX, USA, 6–9 May 1991; pp. 231–238.

12. Nielsen, P. Local approximations: A new way of dealing with irregular waves. In Proceedings of the 20th International Conference on Coastal Engineering, Taipei, Taiwan, 9–14 November 1986; pp. 633–646.

13. Bergan, P.O.; Torum, A.; Tratteberg, A. Wave measurements by a pressure type wave gauge. In Proceedings of the 11th International Conference on Coastal Engineering, London, UK, September 1968; pp. 19–29.

14. Jahne, B.; Klinke, J.; Waas, S. Imaging of short ocean wind waves: A critical theoretical review. *J. Opt. Soc. Am.* **1994**, *11*, 2197–2209. [CrossRef]

15. Zhang, X.; Dabiri, D.; Gharib, M. A novel technique for free-surface elevation mapping. *Phys. Fluids* **1994**, *6*, S11. [CrossRef]

16. Zhang, X.; Cox, C.S. Measuring the two-dimensional structure of a wavy water surface optically: A surface gradient detector. *Exp. Fluids* **1994**, *17*, 17–225. [CrossRef]

17. Zhang, X. An algorithm for calculating water surface elevations from surface gradient image data. *Exp. Fluids* **1996**, *21*, 43–48. [CrossRef]

18. Adrian, R.J. Particle-imaging techniques for experimental fluid mechanics. *Annu. Rev. Fluid Mech.* **1991**, *23*, 261–304. [CrossRef]

19. Wright, W.B.; Budakian, R.; Putterman, S.J. Diffusing light photography of fully developed isotropic ripple turbulence. *Phys. Rev. Lett.* **1996**, *76*, 4528–4531. [CrossRef] [PubMed]

20. Vivanco, F.; Melo, F. Experimental study of surface waves scattering by a single vortex and a vortex dipole. *Phys. Rev.* **2004**, *69*, 1–12. [CrossRef] [PubMed]

21. Grant, I.; Stewart, N.; Padilla-Perez, I.A. Topographical measurements of water waves using the projection moire method. *Appl. Optics* **1990**, *29*, 3981–3983. [CrossRef] [PubMed]

22. Zhang, Q.-C.; Su, X.-Y. An optical measurement of vortex shape at a free surface. *Opt. Laser Technol.* **2002**, *34*, 107–113. [CrossRef]

23. Dabiri, D.; Gharib, M. Simultaneous free-surface deformation and near-surface velocity measurements. *Exp. Fluids* **2001**, *30*, 381–390. [CrossRef]

24. Dabiri, D. On the interaction of a vertical shear layer with a free surface. *J. Fluid Mech.* **2003**, *480*, 217–232. [CrossRef]

25. Tanaka, G.; Okamoto, K.; Madarame, H. Experimental investigation on the interaction between a polymer solution jet and a free surface. *Exp. Fluids* **2000**, *29*, 178–183. [CrossRef]

26. Tsubaki, R.; Fujita, I. Stereoscopic measurement of a fluctuating free surface with discontinuities. *Meas. Sci. Tech.* **2005**, *16*, 1894–1902. [CrossRef]

27. Benetazzo, A. Measurements of short water waves using stereo matched image sequences. *Coast. Eng.* **2006**, *53*, 1013–1032. [CrossRef]

28. Morris, N.J.W.; Kutulakos, K.N. Dynamic refraction stereo. In Proceedings of the 10th IEEE International Conference on Computer Vision, Beijing, China, 17–21 October 2005; pp. 1573–1580.

29. Cochard, S.; Ancey, C. Tracking the free surface of time dependent flows: Image processing for the dam-break problem. *Exp. Fluids* **2008**, *44*, 59–71. [CrossRef]

30. Fouras, A.; Lo Jacono, D.; Sheard, G.J.; Hourigan, K. Measurement of instantaneous velocity and surface topography in the wake of a cylinder at low Reynolds number. *J. Fluid. Struct.* **2008**, *24*, 1271–1277. [CrossRef]

31. Ng, I.; Kumar, V.; Sheard, G.J.; Hourigan, K.; Fouras, A. Experimental study of simultaneous measurement of velocity and surface topography: In the wake of a circular cylinder at low Reynolds number. *Exp. Fluids* **2011**, *50*, 587–595. [CrossRef]

32. Moisy, F.; Rabaud, M.; Salsac, K. A synthetic Schlieren method for the measurement of the topography of a liquid interface. *Exp. Fluids* **2009**, *46*, 1021–1036. [CrossRef]
33. Cobelli, P.J.; Maurel, A.; Pagneux, V.; Petijeans, P. Global measurement of water waves by Fourier transform profilometry. *Exp. Fluids* **2009**, *46*, 1037–1047. [CrossRef]
34. Hering, F.; Leue, C.; Wierzimok, D.; Jähne, B. Particle tracking velocimetry beneath water waves. Part I: Visualization and tracking algorithms. *Exp. Fluids* **1997**, *23*, 472–482. [CrossRef]
35. Hering, F.; Leue, C.; Wierzimok, D.; Jähne, B. Particle tracking velocimetry. Part II: Water waves. *Exp. Fluids* **1998**, *24*, 10–16. [CrossRef]
36. Grue, J.; Kolaas, J. Experimental particle paths and drift velocity in steep waves at finite water depth. *J. Fluid Mech.* **2017**, *810*, R1. [CrossRef]
37. Calvert, R.; Whittaker, C.; Raby, A.; Taylor, P.H.; Borthwick, A.G.L.; van den Bremer, T.S. Laboratory study of the wave-induced mean flow and set-down in unidirectional surface gravity wave packets on finite water depth. *Phys. Rev. Fluids* **2019**, *4*, 114801. [CrossRef]
38. Henry, D.; Thomas, G.P. Prediction of the free-surface elevation for rotational water waves using the recovery of pressure at the bed. *Phil. Trans.* **2018**, *376*, 1–21. [CrossRef]
39. Paprota, M.; Sulisz, W.; Reda, A. Experimental study of wave-induced mass transport. *J. Hydraul. Res.* **2016**, *54*, 423–434. [CrossRef]
40. Brocher, E.; Makhsud, A. New look at the screech tone mechanism of under expanded jets. *Eur. J. Mech. B-Fluid.* **1997**, *16*, 877–891.
41. Ryu, Y.; Chang, K.; Lim, H.J. Use of bubble image velocimetry for measurement of plunging wave impinging on structure and associated green water. *Meas. Sci. Tech.* **2005**, *16*, 1945–1953. [CrossRef]
42. Jayaratne, R.; Hunt-Raby, A.; Bullock, G.; Bredmose, H. Individual violent overtopping events: New insights. In Proceedings of the 31th International Conference on Coastal Engineering, Hamburg, Germany, 31 August–5 September 2008; pp. 2983–2995.
43. Paprota, M. Experimental study on wave-current structure around a pneumatic breakwater. *J. Hydro-Environ. Res.* **2017**, *17*, 8–17. [CrossRef]
44. Iwagaki, Y.; Sakai, T. Horizontal water particle velocity of finite amplitude waves. In Proceedings of the 12th Conference on Coastal Engineering, Washington, DC, USA, 13–18 September 1970; pp. 309–325.
45. Pereira, F.; Gharib, M.; Dabiri, D.; Modarress, D. Defocusing digital particle image velocimetry: A 3-component 3-dimensional DPIV measurement technique. Application to bubbly flows. *Exp. Fluids* **2000**, *29*, S078–S084. [CrossRef]
46. Hassan, M.S.; Khan, M.M.K.; Rasul, M.G. A study of bubble trajectory and drag co-efficient in water and non-Newtonian fluids. *WSEAS Trans. Fluid Mech.* **2008**, *3*, 261–270.
47. Smutná, K.; Wichterle, K.; Večeř, M. Oscillation frequency of bubbles moving periodically in various liquids. In Proceedings of the 35th International Conference of Slovak Society of Chemical Engineering, Tatranské Matliare, Slovakia, 26–30 May 2008; p. 196.
48. Ramírez de la Torre, R.; Vargas, C.; Centeno, M.; Mendez, R.; Stern, C. Characterization of a bubble curtain for PIV measurements. In *Selected Topics of Computational and Experimental Fluid Mechanics*, 1st ed.; Springer: Cham, Switzerland, 2015; pp. 261–269.
49. Murgana, I.; Buneab, F.; Dan, G. Experimental PIV and LIF characterization of a bubble column flow. *Flow Meas. Instrum.* **2017**, *54*, 224–235. [CrossRef]
50. Thielicke, W.; Stamhuis, E.J. PIVlab towards user-friendly, affordable and accurate digital particle image velocimetry in MATLAB. *J. Open Res. Softw.* **2014**, *2*, 1–10. [CrossRef]

Journal of
Marine Science and Engineering

Article

Improving the Performance of Dynamic Ship Positioning Systems: A Review of Filtering and Estimation Techniques

Denis Selimović [1], Jonatan Lerga [1,2,*], Jasna Prpić-Oršić [3] and Sasa Kenji [4]

[1]   Department of Computer Engineering, University of Rijeka, Vukovarska 58, 51000 Rijeka, Croatia;
      dselimovic@riteh.hr
[2]   Center for Artificial Intelligence and Cybersecurity, University of Rijeka, Radmile Matejcic 2,
      51000 Rijeka, Croatia
[3]   Department of Naval Architecture and Ocean Engineering, University of Rijeka, Vukovarska 58,
      51000 Rijeka, Croatia; jasnapo@riteh.hr
[4]   Faculty of Maritime Sciences, Kobe University, 5-1-1 Fukaeminami-machi, Higashinada-ku,
      Kobe 658-0022, Japan; sasa@maritime.kobe-u.ac.jp
*   Correspondence: jlerga@riteh.hr

Received: 13 February 2020; Accepted: 23 March 2020; Published: 30 March 2020

Abstract: Various operations at sea, such as maintaining a constant ship position and direction, require a complex control system. Under such conditions, the ship needs an efficient positioning technique. Dynamic positioning (DP) systems provide such an application with a combination of the actuators mechanism, analyses of crucial ship variables, and environmental conditions. The natural forces of induced nonlinear waves acting on a ship's hull interfere with the systems. To generate control signals for actuators accurately, sensor measurements should be filtered and processed. Furthermore, for safe and green routing, the forces and moments acting on the ship's hull should be taken into account in terms of their prediction. Thus, the design of such systems takes into account the problem of obtaining data about the directional wave spectra (DWS). Sensor systems individually cannot provide high accuracy and reliability, so their measurements need to be combined and complemented. Techniques based on the recursive Kalman filter (KF) are used for this purpose. When some measurements are unavailable, the estimation procedure should predict them and, based on the comparison of theoretical and measured states, reduce the error variance of the analyzed signals. Different approaches for improving estimation algorithms have evolved over the years with the indication of improvement. This paper gives an overview of the state-of-the-art estimation and filtering techniques for providing optimum estimation states in DP systems.

Keywords: ship; navigation systems; dynamic positioning; signal filtering; signal processing; time-frequency representation; Kalman filter; estimation; directional wave spectra

---

1. Introduction

Today, exploitation activities are mainly done in deep waters far from the coastal area. Furthermore, for easier maintenance and transportation, drilling rigs usually float on water. However, cargo ships cannot be anchored under such conditions and need a complex DP technique to maintain their stationary position and direction relative to a reference point on the sea surface. The required criteria for performing a given task, as well as the surrounding conditions and expectations of behaviors for various natural phenomena together have an impact on designing DP systems and their forms of operations.

The design of such systems largely depends on their purpose. Therefore, it is necessary to identify the environmental conditions in which the ship will navigate and perform operations. Namely, ships are exposed to environmental forces and moments such as wind, waves, and current, which produce motion disturbances. To enhance the accuracy of such systems, proper controller design is required to process measured signals from various sensor devices and calculate the ship dynamic properties. Error minimization could be done with a fine estimation of wave conditions. This is significant because of reducing the possibility of critical situations, both for passengers and cargo on board, as well as providing a sense of stability and safety. In addition to moored buoy, satellite, and radar measurements, which have been in use for many years, a DWS estimation based on measured ship motions has proved to be much easier in terms of maintenance [1–3].

In addition to the control algorithm, it is important to determine the proper filtering algorithm (in terms of reconstruction and subsequent estimation) for processing noisy signal measurements. An efficient approach for noise removal is based on time-frequency (TF) distribution methods. Additionally, adaptive, data-driven filtering methods in the TF domain may be used to enhance the accuracy of positioning systems and ship mathematical models. Fundamental methods of signal processing procedures, such as the Fourier transform, assume that the frequency features do not change with time. However, when the spectral characteristics of the analyzed signal are time varying, this approach gives a mean spectral representation and results in low-frequency resolution if estimates are obtained using short time windows or filter banks. Consequently, most of these techniques are inefficient for non-stationary signals, which most often appear in real life. This led to the creation of new concepts for representing signals together in the time and frequency domain [4–6].

Individually, sensors cannot provide enough accurate and reliable information on measured data. Therefore, measurement systems are observed simultaneously, forming an integrated navigation system. Advanced estimation techniques use multiple various sensors together [7]. Missing information from one of the sensors is provided by measurements of other sensors. By comparing the theoretical and measured conditions, it is necessary to reduce the error variation and thus contribute to the accuracy of the system. The use of KF methods has proven to be a very reliable technique in such situations, providing optimal state estimates, both for linear and nonlinear systems.

Techniques to improve the accuracy and reliability of the ship's positioning system have been developed over the years. This paper provides a detailed overview of the methods used in control and filtration algorithms and is structured as follows. Firstly, in Section 2, ship shifts, speeds, and accelerations are considered in terms of their changes during various sea states, as well as the impacts of external forces. The six degrees of ship's motion are divided into three rotational degrees of motion and three linear motion degrees with associated transverse, longitudinal, and vertical accelerations. Next, Section 3 provides a schematic diagram of the dynamic ship positioning system in a closed feedback loop. The mechanical and electrical aspects of the individual system blocks, which together perform the tasks of filtration and signal processing, and finally, the control of the actuators, are described. Related to that, Section 4 gives an overview of DWS estimation methods, as well as an introduction to various signal filtering and reconstruction methods. The section describes new estimation approaches based on ship motion processing. Descriptions of two formulations of estimation methods in terms of the cost function optimization are given.

Section 4 gives an insight into various past and current trends in the TF signal analysis domain. An overview of the standard and adaptive methods is reviewed and divided into non-parameterized, adaptive, and parameterized TF analysis methods. The section also gives a parallel comparison of the TF distributions of measured signals. Real data ship responses are shown for various ship motions, as well as angular velocity and accelerations. The advantages and disadvantages of individual methods are outlined, and it is emphasized that more different methods should be used to achieve better results. Finally, Section 5 deals with advanced filtering techniques that use the KF, based on a series of measurements in time, as well as statistical information of noise and interference. This filtering method has experienced widespread use in navigation systems for estimating unknown variables that cannot be measured directly, as well as for overcoming white and colored noise from the estimated states using measurements from multiple sensors of different accuracy [8–10].

2. Ship Motions and Accelerations

When a ship sways due to the exciting forces and moments of the waves, its swing produces hydrostatic and hydrodynamic forces. Hydrostatic forces return the body to its initial state and are referred to as return forces. Hydrodynamic forces dampen the resulting oscillation motion. Every ship is subjected to six degrees of motion due to external forces. These forces produce accelerations defined as the rate of change of velocity. Under its propulsion, the ship accelerates or decelerates, due to the influence of stern seas or head seas. More importantly, though, are the accelerations caused by external forces such as the wind, sea, and currents [11]. To explain the ship motions as precisely as possible, we will consider the motion's impact on the cargo carried by the ship.

Under the influence of external forces, the ship is subjected to three rotational degrees of motion. These motions are rolling, pitching, and yawing, termed as "rotational movements" since every ship partially turns around its front and back longitudinal axis when rolling, around its transverse axis when pitching, and around its vertical axis when yawing. Furthermore, there are three linear degrees of motion. These are surging, swaying, and heaving introduced as "linear" since the ship surges along its front and back longitudinal axis, sways to some point along its transverse axis, and heaves up and down along its vertical axis [12]. The described six degrees of motion are illustrated in Figure 1.

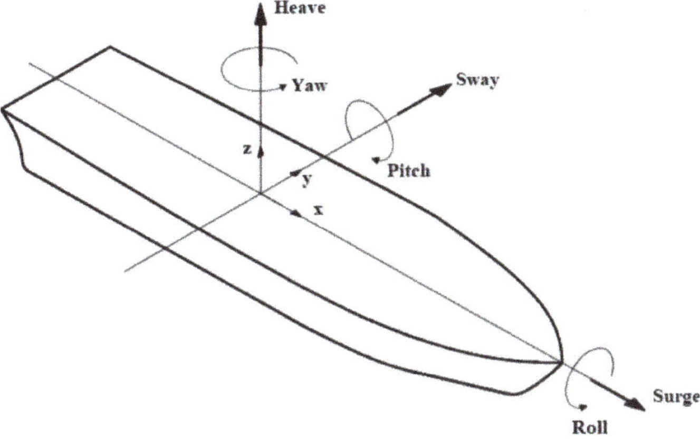

Figure 1. Ship's six degrees of motion due to external forces.

When the ship's cargo encounters the crushing force, the ship lashing system undergoes trials. The ship cargo, which constantly lifts and falls due to heaving, would be lifted and dropped, thereby impacting compression and tensile forces to lashing components. Similarly, when the ship starts moving between two waves, the lashing rods experience forces in opposition to those experienced when the ship moves up a wave profile. Likewise, when ship pitches, the lashing rods tend to be alternately stretched and compressed.

The six degrees of motion combine in different ways to produce the resulting accelerations. First are the transverse accelerations that act across a ship's hull in the transverse direction. Next are the longitudinal accelerations that move along a ship's hull in the front and back direction. The last ones are vertical accelerations that act up and down perpendicular to a ship's hull. Vertical forces are those forces acting normally on the hull, while transverse forces are those acting over it. The acceleration rates are usually estimated for a specific ship before its first operation at sea. Transverse and vertical accelerations are estimated from the shape of the ship surface, its maximum width and length, and the relative positions of the centers of gravity and buoyancy, as recorded from model testing. However, the actual behavior of a ship at sea is not fully known during the ship's design stage, and acceleration values can only be estimated. Rolling, pitching, and heaving are the most common of the six degrees of motion and generate the largest accelerations (pitching and rolling generate acceleration forces that increase with distance from the roll or pitch axis).

When the ship moves through the water with constant speed, the cargo is carried in the same line of motion and with the same speed as a ship. If no external forces act on the ship, then no forces will be transferred to the cargo and associated components. On the other hand, if the ship goes in the direction of some point at the sea surface (by rudder movements or under the influence of heavy seas), there would be an automatic speed reduction producing decelerations that, in turn, would produce certain forces. These forces will be transferred to the cargo and lashing components. When the ship is forced upwards, it eventually comes to rest on the crest of a wave, and when it moves down, it eventually comes to rest in the trough of a wave. During this cycle, the water plane areas are changing rapidly and affect the ship's buoyancy. Vertical acceleration forces are mostly produced during heaving, but also, to a lesser extent, by pitching and rolling. Longitudinal acceleration forces are mainly caused by pitching and, to a lesser extent, surging and yawing. Transverse acceleration forces represent the most important factor when creating the lashing system for a particular sea plan and season of the year. These accelerations are mainly caused by rolling and, to a lesser extent, by a combination of yaw and sway motions. Roll angles are related to the extent of the transverse accelerations produced. If these values fall outside of the ship parameters, for a particular sea route in heavy weather, containers may befall overboard from high stowage locations. High roll angles often occur in following seas. Whilst following seas assist the speed of the ship, longitudinal and vertical accelerations will be generated [13].

External (environmental) forces of induced waves, winds, and sea currents have an oscillatory (stochastic) nature. These forces are introduced as disruptions, which enter the system and have an impact on the ship motions. Conceptually, these forces can be separated into second-order low-frequency forces and first-order high-frequency forces. Therefore, ship motions are often referred to in the context of slowly varying character low-frequency motions and high-frequency motions where the total ship motions are then equal to the sum of the low-frequency and high-frequency components.

High-frequency (wave-frequency) disturbances are produced by high-frequency, wave pressure-induced forces, which have a frequency content at the sum of the wave frequencies and have a fast oscillating character. These forces are not considered in the control algorithm because of too high-frequency content, but may affect the vibrations in the ship hull. On the other hand, forces with low-frequency content can cause the ship to move horizontally, so they are taken into account as feedback signals in DP systems. In addition, the induced forces of wind and sea currents are taken into account. Only mean components are considered for wind forces until the components caused by the gusts are not measured by the ship's responses. The forces of the sea currents have low-frequency content and may also be significant to consider in DP systems [14,15].

3. Dynamic Ship Positioning

When it is required to keep the fixed position and direction (heading) of the ship or to perform complex manoeuvres from one point to another, with low speed speed and a certain accuracy, the DP systems are used. Positioning is done in three degrees of freedom, surge (forward motion), sway (transverse motion), and yaw (rotation about the vertical axis). The operation of the DP system is based on forming a smooth and feasible reference path described in terms of position, direction, speed, and acceleration. The path is formed by an algorithm that uses the current and the desired ship position/direction. Furthermore, the algorithm takes into account the ship's mathematical model with information on the desired goals, operator decisions, and other user-input parameters [12,16–18].

3.1. System Structure

The DP system consists of a few separated modules that individually perform their tasks and jointly compensate environment interferences, issue actuator commands (allocation of rudders, propellers, and thrusters), and ultimately, perform tasks specified by the operator. The block scheme of the control system divided into the fundamental components is given in Figure 2. The complexity of the system stems from the need to maintain a position with defined accuracy. It is a closed-loop feedback control system in which the desired parameters are compared with the estimated values. By measuring the position of the ship and using the control and filtration algorithms, it is possible to determine the required forces for the actuators, adequate for suppressing the surrounding navigation forces. To enhance accuracy, an open loop with a feedforward signal, generated by the environmental forces block model, is additionally used [18,19].

The task of the guidance system is to plan and generate a trajectory path from the start to the desired point. For this purpose, the operator needs to set desired trajectories in terms of endpoint and parameter selection, such as speed and radius of rotation. The control system processes motion-related signals to figure out the ship's state and generate a suitable command for the actuators. Usually, with low-speed applications, the desired parameters calculated by the controller are the surge and the sway forces, as well as the yaw moment [20]. This system has several operation modes that are selected depending on the desired task, such as path tracking, positioning, and damping of roll-motion and pitch-motion. Depending on the mode selected, the control element calculates the required thrust, taking into account the current position and direction of the ship, the environmental conditions, and other operator-specified parameters [12,17].

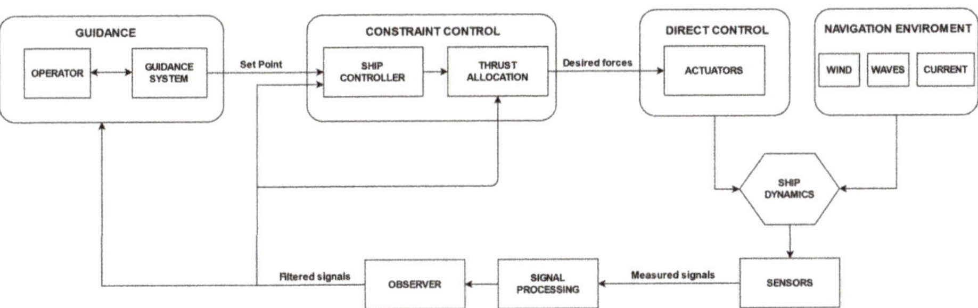

Figure 2. Components of a dynamic ship positioning system.

The direct control block is based on the performance of the actuators and performs the commands of the control system. This block is formed of generator and propulsion elements. Diesel generator power, along with the control signal, is transmitted to an electromotor of the propulsion element via a switchboard. The electrical energy is further converted into mechanical through a shaft that drives the propellers and thrusters. The control signal was previously specified by the control block and carries information on the speed and direction of the thrust [17,18].

Mapping of desired forces and yaw moment, from the controller to propeller system settings (rotational speed, pitch ratio, the angles of the rudder blade, and azimuth thrusters), is done using the thrust allocation system. With the optimization of these settings, power consumption can be significantly reduced as the control takes place in a closed loop with the error minimization process [15,19,20]. Namely, DP systems consume a large amount of energy, and it is important to control them as efficiently as possible. Given the operational conditions of ships, it is also significant to consider the proper maintenance of such systems, which can be achieved by using conventional condition monitoring and condition-based maintenance technologies. To obtain the efficiency of the entire DP system, besides the deviation of the position and heading, it is necessary to take into account the energy efficiency, as well as the influence of the surrounding conditions. The operational efficiency can be expressed by the appropriate key performance indicator defined as a function of the control deviation, energy consumption, and environmental conditions while adjusting the normalization factor. This way, it is possible to perform system maintenance promptly and improve the efficiency [17].

In addition to easier maintenance and enhanced safety, it can be concluded that DP systems also have some disadvantages regarding the complexity and sensitivity of the operation of individual components. These can be power or actuator breakdowns, as well as problems with measuring and control devices. Ships equipped with DP systems require a larger number of thrusters, therefore increasing the electrical load. Electrical systems that supply other DP system components generally use a closed bus configuration, powering all loads from a single bus. Such a configuration reduces the emission of harmful gases and increases the efficiency of the whole system. On the other hand, there is the possibility of losing energy on all components in the case of failure. To increase reliability, the system can be designed by configuring separate buses. However, this configuration has shortcomings in terms of efficiency, so investigators are still turning to closed buses and trying to solve fault problems (such as short circuits) to make closed bus configurations more reliable [12]. Furthermore, the question of control system robustness is important because of a nonlinear mathematical model of the ship's motion. Hence, to achieve good efficiency and reliability, the proper design of the controller should be considered.

3.2. Signal Filtration and Reconstruction

The position and direction of the ship are time-dependent variables, and the values are based on collecting information from different sensors. These variables depend on the operation of the actuator and the surrounding conditions. The ship's position is usually measured by a differential global positioning system (GPS), while the direction is measured with the gyrocompass. Mostly, ships contain more than one gyrocompass and GPS device [16]. To enhance system reliability, additional sensors are often used. The requirements of Global Navigation Satellite Systems (GNSS) in maritime activities are not yet clearly defined, but the worldwide coverage of such systems is very high and of satisfactory quality in terms of position measurement. Therefore, GPS satellite systems are mainly used to measure the ship position. More advanced GNSS devices such as the Russian GLONASS, European Galileo, and satellite-based augmentation systems appear very rarely. The development of combined GNSS receivers is expected in the future. To determine the most accurate position data, all vulnerabilities of maritime GNSS standards should be taken into account and developed according to aviation standards [21]. Inertial navigation system (INS) sensors such as an accelerometer that measures acceleration and a gyroscope that measures the angular rotation of a reference object, as well as Doppler navigation systems are also used [22].

Each of the above-mentioned navigation systems has different characteristics and individually cannot deliver high accuracy and reliability. For this reason, a combination of multi-sensor measurements is used to create the so-called integrated navigation system, which is the subject of many current kinds of research in maritime technology. Namely, if required measurements are not available at some point, the observer should be able to predict the movements of the ship based on noised measurements. Missing parameter values should be estimated from a mathematical model that provides a connection between estimated and measured values. The adjustment of the model is based on tuning during sea trials, and with more measurements obtained by the sensors, the accuracy of control could be greater. In this way, the adapted ship model is increasingly approaching the real one. For example, Fu et al. (2019) [7] in their paper proposed such an integrated system algorithm. The algorithm proved to be suitable for exploiting the characteristics of all measurement systems and to perform a satisfactory evaluation of the results in cases where the GPS did not provide enough data. In this case, a combination of an inertial navigation system and earlier ship speed data was used to evaluate GPS data. The INS is very important because of the autonomy property. Furthermore, in such situations, the aforementioned Doppler navigation system, which measures the speed of the ship, is used as an auxiliary system independent of external data. The main idea behind such systems is to compensate for measurement errors from certain sensors.

Figure 3 represents a block diagram of a structure in a multi-sensory environment. In the concept presented, the INS system uses inertial sensors to calculate the position, velocity, and attitude. The limitations of these sensors can be compensated by using some other types of sensors. GPS sensors also have their disadvantages in terms of signal loss and difficult operation in severe conditions. Therefore, to evaluate the state of nonlinear systems in a noisy environment, highly important signal processing and filtration need to be done before transmitting a signal to the controller.

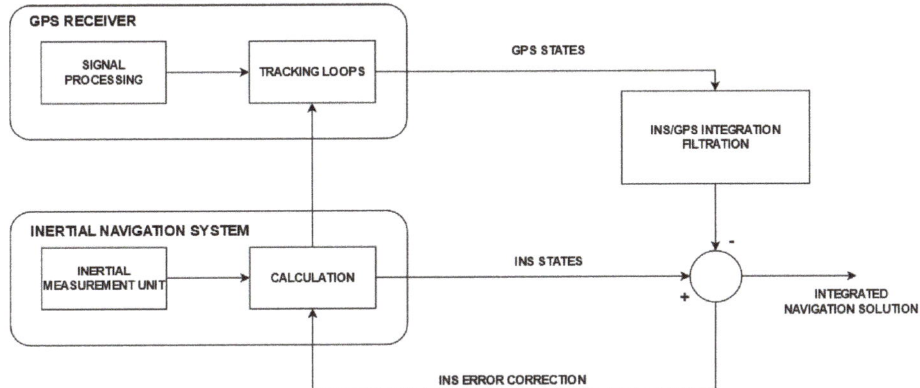

Figure 3. Conceptual design of the integrated navigation system with INS and GPS sensors.

The DP system observer controls measured signal values from various sensors and performs quality analyses to detect interferences and errors to filter out signals before their usage for the control. In other words, the observer's task is to filter out the rapidly oscillating movements caused by the disturbances. Filtration needs to provide only useful ship or sea response information for producing accurate data about the position, direction, velocity, and other measurements [9,10]. The DP system uses this information for appropriate control, and these values are determined by the estimator operation. The operation requires the regulation of only low-frequency motions or the sum of the low- and high-frequency components, which is achieved by filtering wave-frequency components from response measurements and estimated speeds [15].

Initially, proportional controllers were used. Using such controllers, the amount of nonlinearity characteristic of the dead band varies depending on the weather conditions and the operator's selected action. This is comparable to wave filtration since no output action will appear if the control signals are out of the dead band. PID controllers are also used, particularly in cascade with low-pass and notch filters, but performance degradation in the case of high gain control and the phase delay characteristic restrict the use of such filters [15]. In their work, Tannuri et al. (2010) [19] highlighted the advantages of the proposed adaptive controller over a conventional PID controller. They emphasized the robustness of such designs to different environmental conditions, as well as easy adjustment.

An alternative to these filter methods is based on the induced forces motion model and filtration of only low-frequency components of forces. Filtration can be done by using adaptive data-driven TF filtering methods. Such adaptive methods use weighting functions, so higher weighting components are related to signal components, and lower weighting components represent the interferences. Furthermore, for signal reconstruction, the inverse of the TF distribution of the signal is taken.

More advanced filtering techniques can also be implemented using KF (discrete KF, continuous KF, and an extended version of KF) methods or so-called nonlinear observer methods, to achieve maximum control accuracy. KF offers the ability to use a combination of measurements with multiple sensors, allowing the estimation of values that are not directly measurable or missing, such as ship speed during slow positioning operations near exploitation platforms. However, KF formulations require the linearization of the nonlinear equations of the ship model and are limited to white noise, which impairs the stability of use. Another method used in positioning navigation systems is the so-called nonlinear observer suggested by Fossen and Strand (2001) [23]. The nonlinear observer proved to be stable and effective in terms of convergence, as, for example, confirmed by the simulation results by Lie et al. (2019) [10]. An overview of the fundamental and adaptive TF distribution methods, as well as advanced filtering methods based on KF is given in Sections 5 and 6.

4. Estimation of DWS

Hydrodynamic wave-induced ship motions can affect all ship systems, as well as the ship's crew. By predicting the forces and moments acting on the hull, it is possible to predict the movements of the ship in particular conditions, even in real time. Improving accuracy and optimizing the DP system performance is done with a fine estimation of DWS, which further improves the efficiency, safety, and stability of navigation, reflecting on the ship's system, as well as the crew. With available measurements, the fundamental problem is to get information about DWS data to update the system continuously.

Conventional techniques utilize moored buoys for measurements of the wave's height and direction, as well as satellite and radar devices. Moored buoys are reliable and are the primary means of collecting wave data. This issue of such devices exists because of the performance and maintenance complexity, low-resolution properties, as well as their price and high sensitivity to system errors and weather damage [1–3,11,24]. Due to the disadvantages of conventional devices, DWS estimation methods have been extended to process measurements of ship motions (especially the roll, pitch, and heave motions) and their couplings. Such procedures can be done on-board and do not require complex devices. Furthermore, measurements are much easier to implement and provide accurate and reliable outcomes. According to the wave buoy analogy, a ship can also be observed as a unique buoy (sensor). Therefore, the ship can represent a hydrodynamic response to the waves and provide information on the state of the sea, using all actual sensors in real time. The similarity with the process of moored buoys is in the assumption of a linear relationship between waves and associated motions. This approach is still under development and could prove to be a very good alternative to conventional estimation methods. Figure 4 represents a block diagram of the principal idea of the DWS estimation process with the use of measured motion responses. One can see that the purpose of this technique is to perform an estimation of input data based on measured output data. After signal processing, the measured and calculated response spectra are compared with each other, and the measurement error can be minimized to give a more precise and reliable estimation. From the 1970s to the present, many studies have been conducted, and many methodologies have been developed concerning the estimation of DWS using motion responses. Research on the estimation procedures up to 2010 was mainly based on the wave buoy analogy with formulations in the frequency domain. With the improvement of computing resources, nonlinear optimization methods [2] were further developed, as well as formulations in the time domain [25]. A full review of common methods and formulations was given in the work of Nielsen et al. (2017) [1].

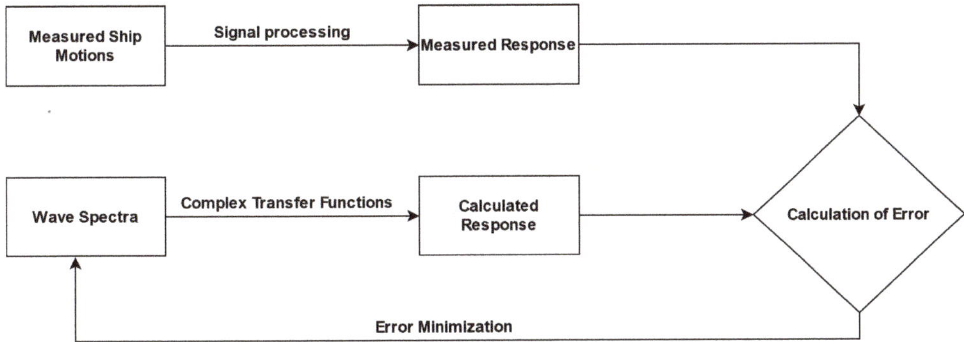

Figure 4. Block diagram of the Directional wave spectra (DWS) estimation method by using measured ship motions.

Estimation methods depend on the previous knowledge of the ship performance in particular wave conditions. These characteristics are specified using complex transfer functions (a complex number that contains the amplitude and phase information) or so-called response amplitude operators (RAO), also indicated in the block diagram in Figure 4. Assuming linearity, the mathematical relationship between the measured response spectra and the DWS in the frequency domain is theoretically given by the following equation [1,24,26,27]:

$$S_{ij}(\omega_e) = \int_{-\pi}^{\pi} H_i(\omega_e, \theta)\overline{H_j(\omega_e, \theta)}E(\omega_e, \theta)d\theta \tag{1}$$

where the theoretical relationship of the ith and jth components of the measured response spectrum $S_{ij}(\omega_e)$ and the directional wave spectrum $E(\omega_e, \theta)$ is given by the complex transfer functions H_i and H_j (complex conjugate). Herewith, the DWS and measured response spectrum are given for encountered frequencies ω_e, wave frequencies measured from the ship's point of view. A graphical representation of the DWS estimation for the formulations in the frequency domain is given in Figure 5.

Figure 5. Illustration of the main idea of DWS estimation formulated in the frequency domain.

RAO transfer functions and DWS are given for different wave directions relative to some fixed direction θ. The representation example of RAO amplitudes, relative to the wave frequency and wave direction, are given in Figure 6 for three ship responses, roll, pitch, and heave. Amplitudes are shown for a ship speed of 16.19 knots and various headings. To improve the view, extra points are obtained by the interpolation process. The example data were derived from the data of a 28,000 DWT bulk carrier, 160 m in length, which sailed in the southern seas.

Considering the ship course (which we can denote by α) as a fixed direction, the wave direction will be equal to the heading of the ship relative to the waves (the so-called encountered angle, which can be denoted as β). This equality follows from the following equation [11,27,28]:

$$\beta = \theta - \alpha \tag{2}$$

This encounter angle corresponds to head waves if $\beta = \pi$, and the relation for all sea states is shown in Figure 7, with β given in degrees.

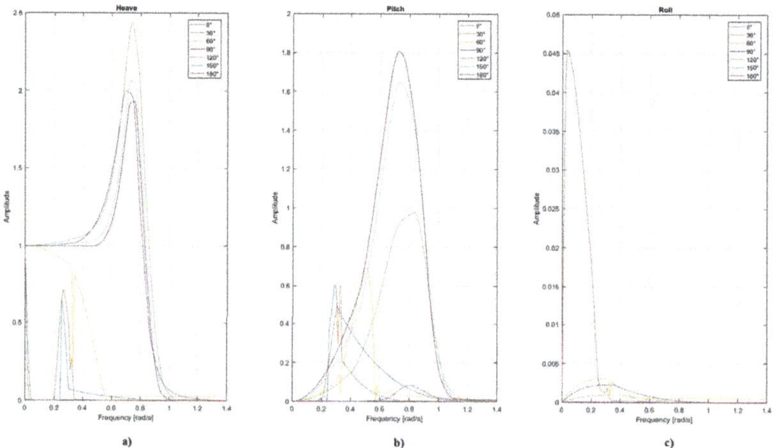

Figure 6. Amplitudes of Response amplitude operator (RAO) for (**a**) heave, (**b**) pitch, and (**c**) roll responses.

The mathematical model of the DP system is formulated to compare the measured (left equivalent in Equation (1) and theoretical (calculated right equivalent in Equation (1)) spectral responses, also shown in the block diagram in Figure 4. Typically, three independent motions are simultaneously considered to uniquely determine the wave direction reliably. These are the roll, pitch, and heave motions that are least caused by the thrusters [24]. A direct comparison of the theoretical relation can be obtained by defining governing equations of the estimation problem with a set of measured spectral responses. This problem can be solved by the minimization process defined with [1]:

$$\theta^2(x) \equiv ||Af(x) - b||^2 \tag{3}$$

where $f(x)$ is a vector function of DWS and b contains information about the measured response spectrum. The A matrix is designed based on the RAO transfer functions.

The above theoretical Equation (1) assumes zero-forward speed, but since the wave spectrum is favorably estimated in the domain of true wave frequencies (ω), it is necessary to consider this forward speed, which further contributes to the complexity of the system [24]. Namely, under the condition of following seas (direction shown in Figure 7), at specific encounter frequencies ω_e, three real wave frequencies are of great importance in the calculated spectrum. Therefore, the connection between encounter and true-wave frequencies is described as a triple-valued function. In other words, the problem exists in the transformation of encountered frequencies into the true-wave frequencies because of the Doppler effect (also known as the triple-valued function problem or the speed-of-advance problem) [3]. The deep-water relationship between the encounter and true wave frequency is given by:

$$\omega_e = \omega - \omega^2 A, A = \frac{U}{g}\cos\theta \tag{4}$$

where U is the ship forward speed and g is the gravitational acceleration. The problem arises when $\omega_e < 1/4A$ because the three-wave frequencies correspond to one positive encounter frequency [1]. Considering the forward speed and observing the relationship between the encounter frequency and the wave frequency for certain conditions in following seas, it is confirmed with the successful results by Nielsen and Iseki (2012) [29], Nielsen et al. (2013) [30], as well as by Montazeri et al. (2016) [31]).

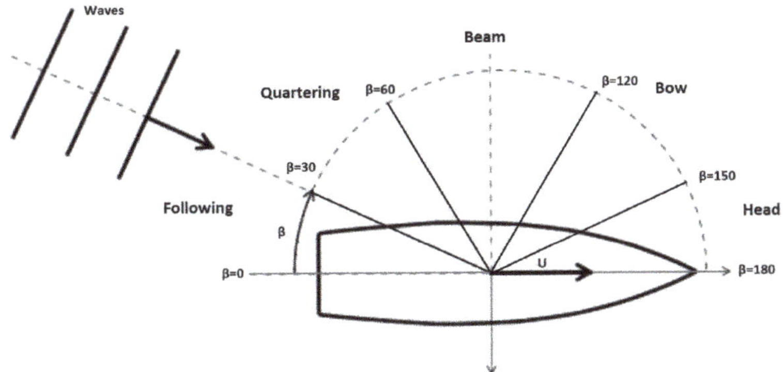

Figure 7. The relation of encounter angles β for different sea conditions.

The above-mentioned minimization problem can be solved using two concepts, one with the parametric formulation and one with the non-parametric (Bayesian modeling) formulation. With the non-parametric formulation approach, DWS is calculated directly without any assumption about the directional wave spectrum. Such formulations are based on the use of the Bayesian technique in iterative schemes for finding components of DWS. The parametric formulation was created in much the same way using the same cost functions. Its methodology assumes that the spectrum consists of the parameterized parts of wave spectra (spectral shapes with mathematical representation and unknown coefficients) that are summed up. In the proposed methods, the most commonly accepted parametric form was to use the summation of the JONSWAP spectrum, each of which directed the cosine by the spreading function [2,24]. The block diagram in Figure 4 practically gives an insight into the similarity of the two discussed formulations. Namely, both concepts are based on the linear spectral analysis to define the conditions that connect the measured ship responses and wave energy spectra. The two formulations of the parametric and non-parametric approaches have been compared in many studies based on simulations and real data, as well as with and without forward speed. Both approaches have been shown to have their advantages and disadvantages and should be viewed as comparative [30].

Non-parametric formulations with Bayesian modeling have mainly been developed to solve the stated triple-valued function problem in the frequency domain. These methods were largely addressed by Iseki (2009) who continued the research on DWS estimation using non-stationary ship motion data [32] with an optimized real-time Bayesian algorithm based on iterative calculations of nonlinear equations. This algorithm has been proven to be satisfying, but iterative calculations require much CPU time, which is a problem for real-time DWS estimation. Pascoal and Guedes Soares (2008, 2009) [33,34] were also concerned with the real-time estimation of wave spectra. Their work was based on the comparison of the nonlinear optimization algorithm with parametric and non-parametric concepts, as well as the KF-based spectral estimation algorithm. Further, validation of algorithms based on experimental data was given later by Pascoal et al. (2017) [2]. This paper describes two types of algorithms for real-time wave-spectra estimation from ship motions. The first algorithm deals with the nonlinear optimization of the spectral model, and the second observer is based on KF. It was concluded that the method is reliable for small ships while the larger ones should be using a fusion of various sensors and take into account some other ship motions. In other words, taking into account higher frequencies cannot be observed with these methods. For this purpose, it would be helpful to use Kalman's filtering methods to accomplish a faster real-time estimation. The lack of this method is evident in the absence of a smoothing process, which can sequentially lead to an inaccurate estimate [2,34]. However, comparing this algorithm with algorithms of nonlinear optimization, it was concluded that, regardless of the smoothness, there is a possibility of spectral estimation achievement similar to the parametric concept of estimation.

The above-described methods for sea state estimation in the frequency domain are based on a direct comparison of measured and theoretically calculated response spectra, with Bayesian modeling or with parameter optimization of the parameterized wave spectrum. There is another way based on the equivalence of energy with the integrated variant of the above Equation (1), which focuses on spectral moments. A detailed description of the technique was provided by Montazeri et al. (2016) [31], as well as by Nielsen (2018a) [35]. The main idea was to create a robust process that would increase efficiency. A further improvement in precision and accuracy based on the buoyancy analogy was confirmed in the study of Nielsen (2019) [3] where multiple available measurements from multiple different ships were used simultaneously. The study was confirmed only on numerical simulations, and it is desirable to extend this study for full-scale measurements.

Another way of sea state estimation with the moored buoy analogy was presented by Nielsen (2018b) [36]. The described simple process relies on direct brute force estimation providing computational efficiency, very convenient to use in real time. It was shown that the wave spectrum estimation can be obtained in a span of only a few seconds, achieving a great cut-off period of the system. Similarly, De Souza et al. (2018) [37] analyzed the use of wave probes on the hull as new degrees of freedom in DP systems by extending the Bayesian estimation method. Experimental measurements were done, and improvement in the estimation over the full range of the expected spectrum was confirmed.

All in all, if this procedure in the frequency domain is used, consistent wave estimates can be expected, but the accuracy mainly depends on RAO reliability. RAO functions are almost always shown to be a weak point due to accuracy or incomplete knowledge of the input conditions. Accuracy also depends on performing a spectral analysis that requires stable operating conditions. Namely, weakness can be generated by variations in the sea state (stochastic nature of waves) and the ship's speed and direction. Signal processing can be done with FFT spectral analysis or by using multivariate autoregressive procedures, depending on the need. There should be a certain minimum period to perform accurate spectral analysis, which is further compounded by the fact that the estimation is not done in real time.

The disadvantages of frequency domain analysis have driven studies in the time domain directly, with an emphasis on better adaptation to non-stationary conditions. A graphical representation of the DWS estimation for formulations directly in the time domain is given in Figure 8. The main idea of this formulation is to obtain real-time sea states from continuous response measurements without taking into account previous measurements and avoiding the use of spectral analysis.

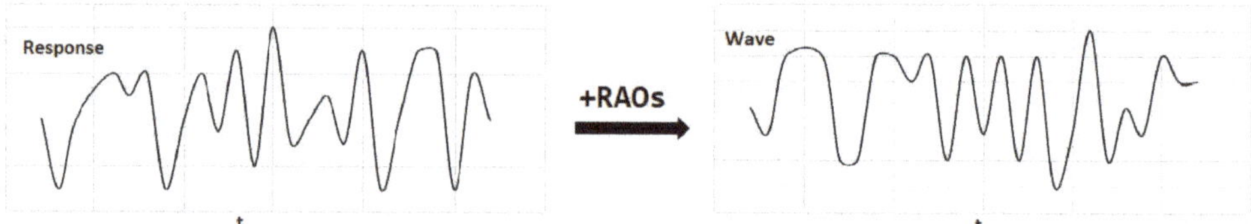

Figure 8. Illustration of the main idea of DWS estimation formulated in the time domain.

So far, several papers have been published on this topic, and the methods are still dependent on the accuracy of the RAO functions as in the case of frequency domain formulations. Pascoal and Soares (2009) [34] proposed a method using the Kalman filter, defining equations in the time domain with the essential assumption that the estimation accuracy is based on the accuracy of the available RAO functions. Furthermore, Pascoal (2017) [2] took into account forward speed, and Ding (2016) [11] validated the results on full-scale measurements, but only for long-crested waves.

On the other hand, Nielsen et al. (2016) [25] proposed and described in detail a stepwise estimation procedure consisting of two steps. The first is a signal-based step that does not require RAO transfer functions for calculation, and the second is a model-based step that uses RAO transfer functions. This is a step forward in terms of eliminating the need to use RAO functions in practical applications due to their unreliability. This method has not been validated and tested for full-scale measurements (experimental data only) and does not take into account forward speed, so its application is questionable. Given the limitation of the method for regular waves only, the proper estimation of irregular waves could be achieved by suitable filtering of the harmonic components of the wave spectrum and using this estimator in future work. Furthermore, by considering multiple responses at the same time, it is possible to estimate the wave direction, which is essential for dynamic positioning systems.

All described DWS estimation techniques can serve as a basis for further research and improvement. Table 1 gives a brief overview of all the techniques of sea state estimation by the wave buoy analogy. The table gives the advantages and disadvantages, both for formulations in the frequency and time domains. We can conclude that an effective sea state estimation can contribute to the safety of navigation and provide an opportunity for easier analysis of ship responses and wave conditions. This could reduce accidents caused by dangerous weather conditions. Overall, this estimation technique may have problems with the accuracy of real-time estimation due to nonstationary data, so further research on the formulation in the time domain is certainly needed to produce a better and more reliable results.

Table 1. Comparison of wave spectrum estimation formulations using the wave buoy analogy.

Formulation	Methods	Characteristics
Frequency domain	Spectral energy distribution Spectral moments	Reasonable estimates of the wave spectrum can be obtained. Estimation accuracy depends on the RAOs and the spectral analysis process. Stationary working conditions required. Need to consider past measurement period, not strictly real-time estimates.
Time Domain	Kalman filtering Stepwise procedure	The past measurement period is not needed. Real-time sea state updates. Lack of tests for real ocean waves. The stepwise procedure partially eliminates the requirements for RAOs.

5. TF Distribution Methods

Classical signal analysis methods, such as the Fourier transform, provide information on signals' frequency content and spectral magnitude assuming that its frequency characteristics do not change with time. For deterministic signals, the instantaneous power analysis and energy density spectrum are used, and for stochastic signals, the analysis of the auto-correlation function in the time domain and its Fourier transform with the power spectrum are commonly utilized. However, it has been shown that most of these techniques are efficient in the case of stationary signals. On the other hand, most of the real-life phenomena exhibit some level of non-stationarity and time-varying spectral features [38].

Non-stationary signals are signals with frequency content varying over time, and a simple example of such a signal is a signal with linear frequency modulation (e.g., a chirp signal). The phase component in Fourier transform keeps information about frequency changes, but this is unsuitable for directly interpreting due to, for example, phase unwrapping. In other words, the signal spectrum obtained by the Fourier transform is a function obtained by separating the signal into groups of wavelengths or complex exponents of infinite time duration (not localized in time). Thus, the Fourier spectrum provides information on the frequencies existing in the signal without their time localization (without time instants of their appearing).

Applying classical techniques for analyzing non-stationary signals results in a low-frequency resolution, especially when utilizing short-time windows. Thus, enhanced frequency and/or time resolution may be obtained by using a proper window length. Its calculation is a challenging task in many practical applications since the optimal window length and type is often data-dependent. This led to the creation of new concepts with simultaneous signal representation in the time and frequency domain. Such concepts are an adequate solution for real-life data processing, so their application can be found in dynamic ship positioning systems.

The following subsections will give an overview of the available TF distribution methods, from linear atomic decomposition distributions, bi-linear (energy) TF distributions, to adaptive and parameterized TF distribution methods. A suitable TF distribution should provide a high resolution and concentration in the time and frequency domain, supporting easier analysis and interpretations of multi-component signals. Furthermore, cross-terms need to be eliminated to separate the real components and noise components. For use in real-time applications, mathematical properties need to be achieved in terms of marginality and non-negativity (and many other properties), and consideration should be given to reducing computational complexity to ensure a satisfactory representation time [39].

5.1. Non-Parameterized TF Distribution Methods

Standard TF distributions belong to the so-called group of non-parameterized TF distribution methods with time and frequency resolution being highly window dependent. The term non-parameterized does not mean these TF distributions are parameterless, but rather that they do not require additional parameters to regulate the shape of the TF cell. With regards to the origin, non-parametrized TF methods can be divided into linear and nonlinear methods. To detect characteristics, the analyzed signal needs to be decomposed on signal components. This process is known as the blind source separation procedure and can be troublesome without any prior knowledge about the signal, its components, and their mixture [6].

5.1.1. Linear Distributions

Linear TF distributions are also called atomic decomposition methods as they separate the signal into weighted sums of components localized in the time and frequency domain. Here, we mention the short-time Fourier transform (STFT) and continuous wavelet transform (CWT) methods. Such methods cannot simultaneously obtain good localization in the time domain and resolution in the frequency domain, so the main idea is to find a suitable method that gives both properties satisfactorily.

A natural extension of the Fourier transform for non-stationary signals is the linear TF transform, called the STFT method, defined as:

$$STFT(t, f) = \int_{-\infty}^{\infty} s(\tau)h^*(\tau - t)e^{-j2\pi f\tau}d\tau \tag{5}$$

where $s(\tau)$ is a considered signal and $h(t)$ is a fixed size window function centered at time instant t. Thus, the time element was introduced into the Fourier transform by "pre-windowing" of the signal in the vicinity of t, and the Fourier transform was calculated locally inside the time window [5].

The length and shape of the window function represent a significant part since they directly affect the resolution in the time and frequency domain. Namely, short time window lengths allow good localization in the time domain. However, due to the uncertainty product, this results in decreased frequency resolution. On the other hand, large time window lengths result in higher frequency resolution and lower time localization. Therefore, it is not possible to obtain good time and frequency resolution at the same time as manifested by Heisenberg–Gabor's inequality, well known from quantum mechanics [38]. Therefore, in practical applications, finding the optimal window length is a difficult task since it depends on the relative values of the position and power level. For time localization of high-frequency components, a shorter time window is used, and a longer time window can provide frequency localization of low-frequency components. In practice, this is problematic especially at observed signal pulses. However, due to its simplicity and precision, the STFT method retains its popularity despite the limited resolution [4,40].

The squared module of the STFT gives a real and non-negative signal distribution, well known as a spectrogram. The spectrogram represents the spectral densities and power changes over time. Here, the local power spectrum can be determined from signal components centered at the successive time points of interest, as follows:

$$\rho_{spec}(t,f) = |S(t,f)|^2 = \left| \int_{-\infty}^{\infty} s(\tau)h(t-\tau)e^{-j2\pi f \tau} d\tau \right|^2 \tag{6}$$

where $h(t-\tau)$ is the time-limited analysis window centered at $t = \tau$ and $S(t,f)$ is referred to as the STFT.

CWT is another linear TF transform, which uses wavelet base functions instead of sinusoidal functions. This method is defined with the mother wavelet, and other wavelets are derived by scaling and shifting parameters a and b in the equation:

$$CWT(a,b) = |a|^{-1/2} \int_{-\infty}^{\infty} s(\tau)\psi^*\left(\frac{\tau-b}{a}\right) d\tau \tag{7}$$

where parameter a indicates wavelet dilatation and b represents wavelet translation affecting time localization. $\psi^*(t)$ is the complex conjugate of the analyzed mother wavelet $\psi(t)$ ($a^{-1/2}$ is an energy normalized factor). According to the mathematical criteria in (7), the wavelet energy is finite and equal for various scale values.

With factor scaling, the signal is expanded (less detail is obtained, but the entire duration of the signal is obtained in time) or compressed (more detail, but there is no full signal duration for such a scale factor). This scaling characteristic provides the ability to adapt the window that is appropriate for non-stationary signals [40]. The mother wavelet is the prototype function, and it is defined first. Next, data analysis is done on multiple levels by analyzing the lower frequencies with a wider window and higher frequencies with a narrow window. The resultant frequency resolution is high at lower frequencies and low at higher frequencies. Thus, the method is suitable for the analysis of narrow-band transient signals to improve the frequency resolution of high-frequency components. The representation of the continuous wavelet transform is named the scalogram. The time-scale scalogram is a two-dimensional function that represents the energy distribution of signals for used scales a and locations b [41].

All in all, there needs to be a trade-off between time and frequency resolution (improving the time resolution decreases the frequency resolution and vice versa). A need for improved, high-resolution TF representation also robust to noise led to the development of other representations as described below. Furthermore, these signal decomposition techniques require separate hardware implementations for forward and backward transformations, which can further affect the complexity of the hardware [42].

5.1.2. Energy Distributions

The above-mentioned TF distributions are linear distributions with the signal being separated into fundamental components (atomic decomposition). On the other hand, bi-linear TF distributions (energy distributions) separate the signal into two variables, time and frequency. An attractive TF energy distribution is the Wigner–Ville distribution (WVD). With WVD, signal spectral structures are considered for each time instant, and estimation of the instantaneous spectrum (the energy density) is done. The Fourier transform of the instantaneous autocorrelation function, i.e., the WVD, is a bilinear transform or, in other words, quadratic TF representation, defined as:

$$WVD(t,f) = \int_{-\infty}^{\infty} s\left(t+\frac{\tau}{2}\right) s^*\left(t+\frac{\tau}{2}\right) e^{-j2\pi f \tau} d\tau \tag{8}$$

WVD represents the fundamental TF distribution method for all other energy distributions. It achieves the maximum energy concentration (maximum TF resolution) for linear modulated signals ("chirp" signals), but for multi-component signals, so-called cross-term interferences or artifacts (caused by the quadratic nature of the WVD) are created and limit the possible application of this method [38,40]. Such terms represent large oscillating terms located in the middle of actual signal components and could be larger than them. External cross-terms appear with a combination of each of the two signal components, while internal cross-terms appear due to the nonlinearity of one of the components [4].

Overall, cross-terms and weak TF support complicate the interpretation of the WVD. Therefore, this method is rarely used to analyze the signal structure. However, this method represents the basis for the development of new adaptive methods such as pseudo-WVD and the adaptive Cohen class distributions. On the other hand, the spectrogram also exhibits the cross-terms, but they show up when the signal components are close in the TF plane. Another shortcoming attributed to WVD is that it performs well only for infinite duration signals. For the time-limited signals (which typically occur in real-life data), the simple bi-linear Fourier transform does not provide highly efficient data analysis. Additionally, discrete WVD (due to the sampling according to Nyquist's law) may contain spectral aliasing. To avoid such cases, the analytic signal with the half frequency band of the original signal is mainly used, extending the potential frequency range of the input signal [5].

5.2. Adaptive Non-Parameterized and Parameterized TF Distribution Methods

Attempts to create a simple method that would exhibit great performances in the analysis of complex signals are often discussed as adaptive TF distribution methods. Adaptive methods tend to improve classical non-parametrized TFA methods in terms of a more trustworthy representation of nonstationary signals without cross-terms by adjusting and optimizing the window length. Hence, adaptive methods differ in the mode of selecting the optimal window length. Many of the researchers went in the direction of smoothing the WVD with 2D Gaussian functions since Gaussian window properties allow minimization of the time-bandwidth product. The customization scheme uses rule-based 1D adaptation (for individual parameters of a basis function), the empirical customization (for additional control parameters), basis function parameter discretization, and parameter optimization with the dynamic kernel design.

With slow frequency changes over time, the rule-based window length calculation proved to be effective for optimized measurements over time. In the case of rapid frequency changes, utilizing measurements only in the frequency or together in the time and frequency domain showed better results for improving the STFT. The disadvantages of these methods are found in the optimization procedure done by joining the fast Fourier transform for each point, as well as sensitivity to noise [6].

Opposite these rule-based methods, another adaptive Stockwell transformation or S-transform method uses a multi-scale instead of a sinusoidal function as a basis function, which creates a link between the CWT and STFT methods. This adaptive method provides fine time resolution at high frequencies. Unlike the rules above, S-transform is a group of signal-independent rules. Except for the window length and its adjusting, adaptive TF methods can also be defined by other basic function parameters such as time and frequency shifts. In other words, discretization of the parameters using the empirical scheme allows adjustment of a particular parameter (constant-Q transform, bionic wavelet transform, and others). More information on the previously mentioned adaptive methods can be found in the literature, for example Yang et al. (2019) [6], Boashash (2016) [38] (2016), or Sejdic et al. (2009) [42].

Furthermore, methods can be based on the application of an adequate kernel for cross-term filtering in the ambiguity domain. Adaptive Cohen class distributions introduce a kernel function and achieve the expected characteristics of the TF distribution, including the elimination of cross-terms and high-resolution maintenance. WVD tuning is accomplished by moving the kernel function in the time and frequency domain. In the ambiguity domain, signal component elements (auto-terms) are mostly located around the origin, while the cross-terms are mainly located out of the origin. With the use of a 2D low-pass filter around the origin in the ambiguity domain, it is possible to reduce undesired cross-terms and obtain significant properties. By introducing windows to WVD, the smoothed pseudo Wigner–Ville distribution (SPWVD) implements a signal-independent low-pass filter to suppress cross-terms. Additionally, Lerga and Sučić (2009) [43] provided an improved window selection for SPWVD based on the statistical test, called the relative intersection of confidence rule. In addition to improved resolution and reduced interference, this method is also simple and not expensive in the sense of calculations. However, the problem arises when the distribution is equal to zero in the TF plane because this method becomes a useless analysis window [38,40].

Some other proposed adaptive Cohen class distributions that follow this type of kernel function are Bessel, Zhang–Sato, Born–Jordan, Choi–Williams, Page, Rihaczek, and Zhao–Atlas–Marks distributions. For example, the Choi–Williams distribution is used in many applications with a constant frequency content multi-component signal [6,42]. Another Cohen class distribution is the S-method (SM) derived from the STFT-WVD relationship. With SM, auto-terms come first in terms of conservation, while reducing the appearance of cross-terms:

$$SM_x(t, f) = 2 \int_{-\infty}^{\infty} P(\theta) F_x(t, f + \theta) F_x^*(t, f - \theta) d\theta. \tag{9}$$

By changing the width of the $P(\theta)$ window, it is possible to perform a transformation from a spectrogram to a WVD to reach complete auto-term integration and eliminate cross-terms (to perform smoothing). Unlike the so-called smoothed spectrogram that uses the convolution of two STFTs in the same direction, SM utilizes two STFTs in opposite directions. That way, SM improves the concentration in the TF plane. Furthermore, the method can be extended to higher order TF representations. Using such a method for TF representations retains automatic expressions similar to those of the WVD, but the terms are significantly reduced [38].

These adaptive kernel methods are efficient as they eliminate cross-terms and give a satisfactory resolution. However, the problem of separating auto-terms from cross-terms can arise when these terms overlap. Whether the kernel is dependent or independent of the signal, kernel optimization surely reduces the energy concentration. Moreover, no kernel can guarantee the suppression of all cross-terms [4,39,40,42].

Due to the encountered problems in the representation of multi-component non-stationary signals, standard TF distribution methods are not sufficient in many real-life applications and need to be improved for successful implementation in specific fields of applications. This has led to the development of various advanced, high-resolution, and reduced interference distributions. TF distribution methods that require a signal model and contain extra parameters are called parameterized TF analysis methods. These methods are formed from non-parametric TF methods and provide a better representation of multi-component signals with a complex TF structure by parameterizing the phase function of the basis function using additional parameters and have gained much research interest. Examples of parameterized TF methods that have been explored so far are warped TF methods, chirplet transforms, adaptive parametric analysis based on atomic decomposition methods, and adaptive non-parametric analysis based on empirical mode decomposition and local mean decomposition (such as Hilbert–Huang transform, which directly extracts the local signal properties using numerical approximation and provides fine TF resolution using the Hilbert transform). More detailed information about parametrized TF distribution analysis was summarized in the papers of Feng et al. (2013) [40] and Yang et al. (2019) [6].

5.3. Comparison of TF Distribution Methods

This subsection provides an overview of the benefits and limitations of the above-mentioned TF distribution methods with examples of measured signals in the TF domain.

Table 2 gives a comparative overview of the important properties of TF distribution methods such as the resolution and mathematical properties, the occurrence of cross-terms, and the computational complexity. The table provides a comparison of some of the non-parameterized, adaptive non-parameterized, and parameterized TF distribution methods for easier understanding of the progress in the development of such methods. Compared to the CWT and WVD methods, the spectrogram gives better results without cross-terms and with reduced interferences, as well as better energy distribution preservation. However, the energy concentration obtained by WVD is significantly higher (best clarity). Furthermore, the scalogram adds artifacts at frequencies close to zero or close to the edges. Although spectrograms and scalograms have limitations, they are widely used in practice for the sake of simplicity and fast implementation. On the other hand, WVD and Cohen's class adaptive distributions introduce negative components and distortions. Adaptive and parameterized methods show improvement in characteristics, but at the same time, there is a problem of calculation complexity.

With non-stationary wave patterns, the spectrogram is usually applied to measured free surface heights [5]. For a steadily moving ship, which produces small wave amplitudes, it was shown by Torsvik et al. (2015b) [44] and Pethiyagoda (2016) [45] that the spectrogram consists of two clear linear components. Those are the sliding-frequency component (caused by divergent waves) and the constant-frequency component (caused by transverse waves). However, for real-time data of fast ship motions, it is noted that the spectrogram has additional components in the case of nonlinear waves. The use of the spectrogram method has had great popularity in the field of signal processing and many other areas of science. It is used in many situations due to its simplicity, robustness, easy interpretation, as well as the property of linearity.

The spectrogram provides the possibility of separating the signal on wave components of different frequencies. Initially, the wave analysis was performed by a non-stationary sensor moving vertically to the direction of ship movement. Didenkulova et al. (2013) [46], Sheremet et al. (2013) [47], and Torsvik et al. (2015) [44] conducted experiments using a stationary sensor.

In the study of Pethiyagoda (2018) [48], the influence of the linear dispersion relation, with two already described clean linear components of the spectrogram, was described. Likewise, the influence of fluid flows (Froude's number) on the linear dispersion curve was also described because it directly affected high-intensity regions on the spectrogram. Identification of the measured responses for different rates of the Froude number (F_H), based on the depth or length, was done first. Linear water wave theory was successfully applied with the corresponding idealized mathematical model to demonstrate the extraction of spectrogram features with the small wave amplitudes produced on the infinite-depth water. It was shown that for the finite-depth fluid, two areas on the spectrogram could be recognized. These are areas of subcritical ($F_H < 1$) and areas of supercritical ($F_H > 1$) flows. In both cases, waves have different shapes due to the presence of the transverse and divergent waves for the subcritical flows, and only divergent waves for the supercritical flows. In a representation of the dispersion curve, it should only be taken into account that the sensors travel on the ship at a relative unidimensional speed, and representations should be displayed in such a ratio. The linear dispersion curve could be obtained for $F_H = 0$ for infinite-depth fluids. For finite-depth fluids and subcritical Froude numbers, the curve is similar to the infinite-depth state, while for supercritical numbers, the curve differs due to only one area (divergent waves).

For supercritical flows, it is interesting to observe the wavelength that could be critical for the coastal area due to variability and possible high values. Namely, when passing through semi-sheltered and coastal areas, it is very important to comply with certain rules concerning the shipping speed because waves created by the ship have a significant impact on such areas, as described by Didenkulova (2013) [46]. Furthermore, in the paper of Torsvik et al. (2015) [49], a wave pattern transformation was observed near the coastal area using the STFT and other linear TF distribution methods. It was shown that the amplitude of the wave and its energy were reduced when approaching the coastal area due to significant divergent waves system reduction and the appearance of wave breaking phenomena. The wave breaking process, on the other hand, did not affect the energy of transverse waves (the energy was stable or even increased).

Parameterized TF distribution methods can also be found in shipboard system applications. For example, Yu et al. (2013) [50] studied the parameterized Hilbert–Huang transform method (empirical decomposition formulation) with ship responses in the case of an underwater explosion damaging process. Compared with the STFT and CWT methods, this method showed a more accurate TF representation of the response in such situations. STFT and CWT provided the possibility of improving temporal resolution as previously known.

Table 2. Comparative representation of TF distribution methods.

Group	Method	Properties	Resolution	Cross-Terms	Complexity
Non parametrised	STFT	Non-negativity Lack of marginality	Poor resolution Limited by constant widnow width	Free from cross-terms Reduced interferences	Expensive when finding optimal window length Less expensive for narrow band signals
	CWT	Feature extraction possibility	Limited variable resolution	Free from cross-terms	Faster than STFT Computational cost depends on scale values
	WVD	Feature extraction possibility	Limited variable resolution	Free from cross-terms	Faster than STFT Computational cost depends on scale values
Adaptive non parametrised	Reduced interference distributions	Mostly marginality retained Lack of positivity	Limited variable resolution Depends on the signal components	Suppression of cross-terms	Slower High computational complexity due to optimization
	SPWD	Lack of non-negativity Marginality no strictly valid	Improved resolution Slightly disrupted by smoothing	Reduced cross-terms Smother due to time smoothing	Simple and not expensive calculations
Parametrised	Atomic decomposition	Finer local properties	High TF resolution	Suppressed	High computational complexity signal decomposition optimization
	Hilbert-Huang	Finer local properties	High TF resolution Finer estimation with more samples	Suppressed	High computational complexity Difficult to resolve signal components

The next figures give examples of comparative representations of measured signals obtained from the aforementioned 28,000DWT bulk carrier. The figures represent the examples of time and TF domain representations of real-life ship pitch and roll motion response measurements. Firstly, in Figure 9, the measured pitch and roll motion responses in the time domain are given with the duration of one dataset of 60 min.

Figures 10 and 11 give the representation of the pitch and roll motions in the TF domain using spectrogram, scalogram, and WVD representation methods. Simultaneous representation of the signal distribution in the time and frequency domain obtains a better characteristics extraction of measured signals. By comparing representations in the TF domain, one can see that there is a similarity between distributions. However, each has its advantages and disadvantages as mentioned before. WVD achieves the best clarity and frequency resolution, but for the multi-component signal, like these, it introduces much more interferences. On the other hand, these cross-terms interferences are eliminated on the spectrogram and scalogram representations, but their resolution is worse.

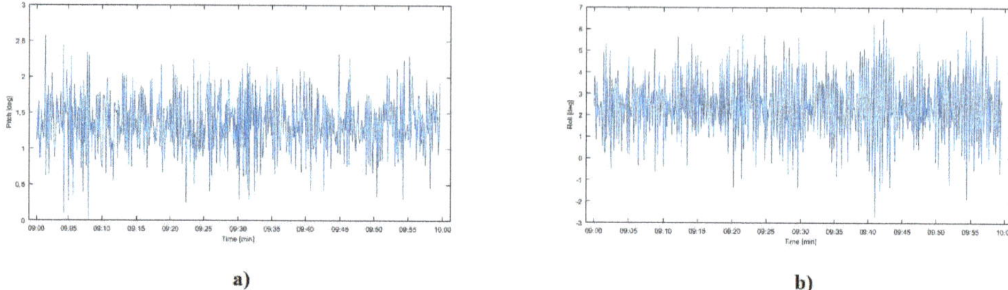

a)   b)

Figure 9. Representation of measured responses in the time domain for (**a**) pitch and (**b**) roll motions.

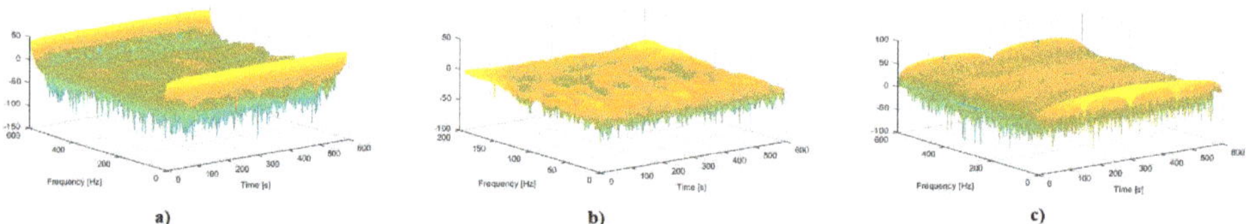

a)   b)   c)

Figure 10. Time-frequency (TF) representations of measured pitch responses: (**a**) Short-time Fourier transform (STFT), (**b**) Continuous wavelet transform (CWT) and (**c**) Wigner-Ville distribution (WVD).

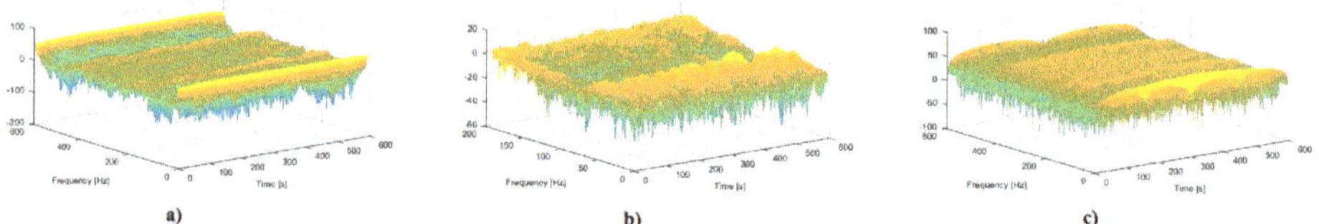

a)   b)   c)

Figure 11. TF representations of measured roll responses: (**a**) STFT, (**b**) CWT and (**c**) WVD.

By combining and adapting existing feature extraction methods of various techniques, it is possible to enhance the accuracy of DP systems. If any of the distributions do not meet the application requirements, then it is necessary to either combine or adapt the existing fundamental methods that are described before and shown as examples in the figures. Adapting the appropriate TF distribution method for use in the signal filtering process can make a scientific contribution in terms of improving the accuracy of the estimation algorithm in dynamic positioning systems. After adequate processing and filtration of the measured motion signals and with knowledge of the complex RAO transfer functions, it is possible to obtain more accurate information on the DWS by combining ship motions for different encounter angles and encounter frequencies.

6. Kalman Filtering Techniques

Position, direction, and speed measurements in navigation systems are based on data collected from various sensors that are sensitive to noise and error. To estimate the information about forces and moments acting on the hull of a ship successfully, it is essential to detect and correct sensor errors, taking into account the measurement models of waves, wind, and sea currents [9,18]. For this purpose, an advanced filtering technique based on KF model can perform the filtration of the oscillatory wave-induced forces from the noisy signal measurements and provide optimum estimated states.

When designing a KF system model, it should be observable to achieve the convergence of the estimated states to the probabilistic expected value. State estimation is done by multiplying the prediction and measurement probability functions together, scaling the result, and computing the mean of the resulting probability density function. Optimum states can be reached with the minimization of the error variation by using a combination of various sensors and a feedback loop [14]. The optimum state, in terms of the minimum error covariance, provides the ability to estimate the state of the dynamic system from the input-output pair in the feedback loop. A block diagram of such a system model is given in Figure 12. The main idea is to control the error between the measured and the estimated values at zero with variable gain. In the model, noisy measurements are taken as a feedback signal based on which the estimation is made.

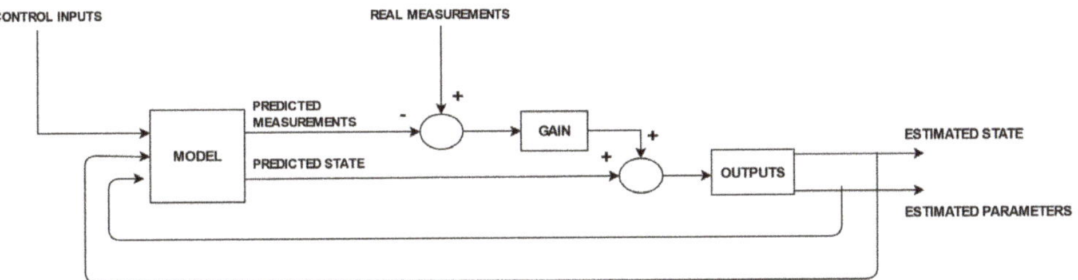

Figure 12. Block schema of the Kalman filter (KF) model.

Computationally, the multiplication of these probability density functions relates to the discrete KF equation designed for stochastic systems, which is similar to the state observer for deterministic systems [14,15,22,51]:

$$x_{k+1} = Ax_k + Bu_k + w_k \tag{10}$$

The state vector x_k contains information about the position, direction, and speed, and these variables should be estimated. It represents the a priori state, while the vector x_{k+1} represents the a posteriori state. The variables to be estimated are given by matrices A (state transition matrix) and B (control matrix). The variable u_k represents the input from which the estimate is derived, and w_k takes into account the noise. KF is a two-step algorithm. It consists of a prediction part (time update equations) and an estimation part (measurement update equations).

Firstly, after measurements are taken, the prediction part is done. The linear system model, without dynamic noise taken into account, is used to calculate the a priori state estimate $\hat{x}_k$ and the error covariance P:

$$\hat{x}_k = Ax_{k-1} + Bu_k, \tag{11}$$

$$P_k^- = AP_{k-1}A^T + Q, \tag{12}$$

The a priori error covariance matrix P is based on knowledge about the difference between measured states and previously estimated states. Matrix Q is the system process noise covariance matrix, defined in time intervals. It collects data about unmeasured dynamics and sensor noise.

The second part of the algorithm uses the a priori estimates calculated in the prediction step and updates (correct) them to find the a posteriori estimates of the state and to minimize error covariance. The state estimate of correction is given by:

$$\hat{x}_k = \hat{x}_k^- + K_k(y_k - H\hat{x}_k), \tag{13}$$

$$P_k = (I - HK)P_k^-, \tag{14}$$

where H is a measurement matrix related to the connection between the current state and measurement. The expression $(y_k - H\hat{x}_k)$ represents the deviation of the actual y_k measurement from the predicted measurement. The Kalman gain K_k is calculated so that it minimizes the a posteriori error covariance:

$$K_k = \frac{P_k^- H^T}{H P_k^- H^T + R} \tag{15}$$

where R is the measurement noise matrix.

Once the update equations are calculated, in the next time step, the posterior estimates are used to predict the new a priori estimates, and the previous steps are repeated. To estimate the current state, the algorithm does not need all past measurements. Only estimated states, the error covariance matrix from the previous time step, and the current measurement are needed. In other words, the KF is a recursive method. However, the recursive form is not the best choice for real-time implementations because of the problem of numerical calculations of covariance matrices. By quadratic filtering, better accuracy and stability can be achieved [14,22]..

The KF is essentially linear, so it is necessary to linearize (approximate) the complex nonlinear equations of the DP system to remove noise. Such calculations represent a strain on the system since the calculation of parameters should be obtained in real time. Furthermore, the KF is driven by white noise, which further contributes to the problems and limitations of achieving an accurate estimation [22,52].

More accurate approximation of nonlinear systems can be done by utilizing the extended Kalman filter (EKF). EKF performs linearization of the nonlinear function around the mean of the current state estimate. This formulation also performs calculations of the Jacobians matrices and tunes the gain while updating the states using a linear function. However, EKF has some drawbacks. There is a problem in the calculation of the Jacobians analytically due to complicated derivatives and the high computational cost to calculate them numerically. Furthermore, a system with a discontinued model is not differentiable, and Jacobians do not exist [51]. To allow the Jacobian's free estimation, Alminde et al. (2007) [53] studied the application of integrated quantized state simulation to propagate the state and obtain a backward variance estimate of the Jacobian and ultimately confirmed their theory. The EKF algorithm proved to be satisfying in numerous marine navigation applications. For example, Kuchler et al. (2011) [54] confirmed the improvement in the estimation of the heave motion using the acceleration measurements and EKF filter. Furthermore, Assaf et al. (2018) [55] in their study extended the use of EKF to discuss more general nonlinear estimation problems. The proposed algorithm was shown to be effective in tracking normal ship speed.

When the linearization procedure does not provide a fine approximation for highly nonlinear systems, another estimation method called the unscented Kalman filter (UKF) is used. Instead of approximating a nonlinear function as EKF does, such filters approximate the probability distribution [9,51,52]. The algorithm is based on Bayesian theory and the deterministic sampling technique, also known as unscented transform (UT). Shown in Figure 13, the filter selects a minimal set of sample points (also referred to as sigma points and symmetrically distributed around the mean) from the Gaussian distribution, and it propagates them through the nonlinear system model. By utilizing the UT transform, it estimates the mean and covariance values of the new transformed sample point dataset and uses them to find the new state estimates. UKF is not robust to a non-Gaussian environment with outliers, and there is a need for a different robust solution. In their study, Peng et al. (2019) [9] proposed modified UKF to reduce the influence of outliers on estimates. In the proposed study, they confirmed the improvement in terms of stability, convergence properties, an accurate estimation of all given states.

Research of Vaernø et al. [8] gave an overview of dynamic algorithms for ship positioning with four different control models, including the linear KF, EKF, and UKF. The results confirmed that the DP process was predominantly linear. Slight improvement in performance can be achieved by using nonlinear damping (namely by including the EKF or UKF algorithms). Linear KF behaves similarly to EKF and UKF only when linear damping is used. The control implementation and observation algorithm largely depend on how well the underlying model embraces the dynamics of the system. Furthermore, a detailed description of the high confidentiality model with various models of controller and observer designs was given by Fossen (2011) [15] and Sorensen (2011) [56].

Another nonlinear state estimator, based on a similar principle, is the non-parametric approximation technique called the particle filter (PF). It also uses sample points referred to as weighted particles. The flow of a set of particles from the prediction phase through the update phase to the resample phase and back to the prediction phase, with a non-parametric representation of the distribution density, is shown in Figure 14.

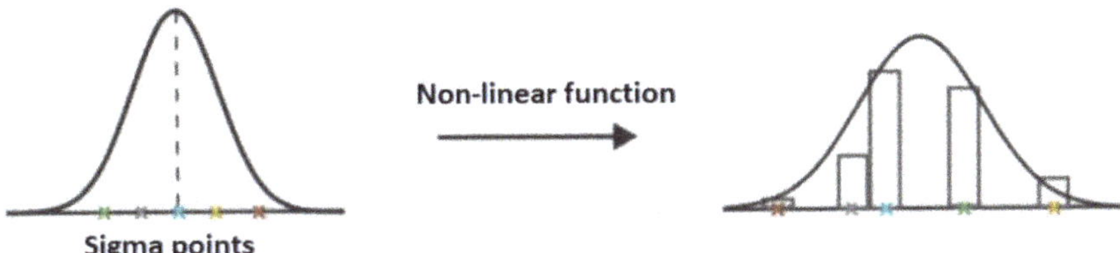

Figure 13. Probability distribution approximation using Unscented Kalman filter (UKF) with sigma points.

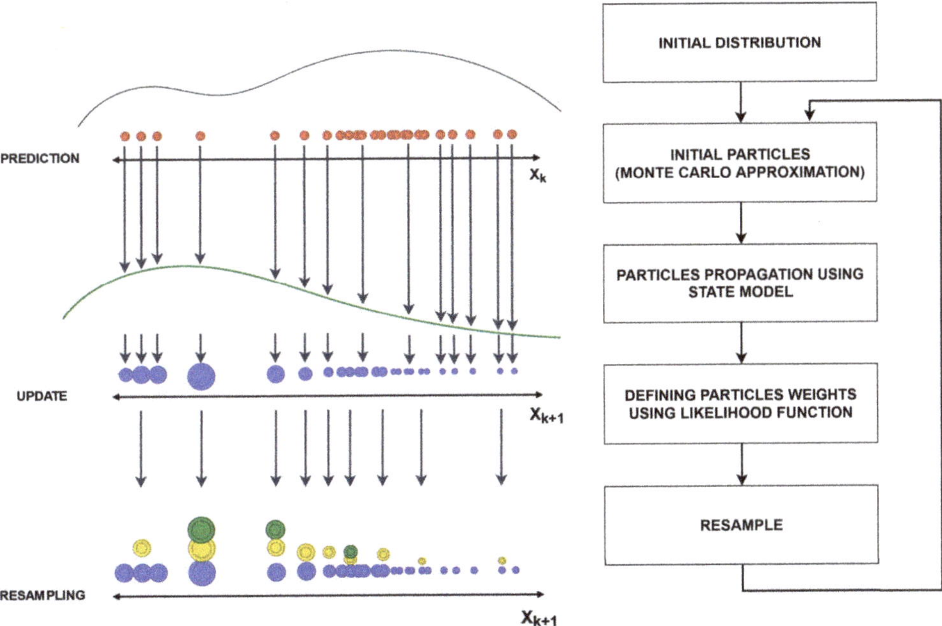

Figure 14. The main principle of particle flow in the Particle filter (PF) algorithm.

This filter can approximate any arbitrary distribution, and it is not limited to a Gaussian noise only. It is based on the Monte Carlo method, and approximation is done using the probability density function with the sample mean instead of the integral operation [52]. The more accurate approximation is obtained by use of more particles. On the other hand, in relation to UKF, it is a computationally expensive filter since it requires a larger number of particles to represent a distribution not known explicitly [57]. Furthermore, the problem of particle sample degradation occurs due to the variation of particle weight, increasing with the iteration time. This leads to inefficient use of resources and the decreased efficiency of approximation. Due to the structural complexity and computational cost, these types of filters are not often used in control systems, but their development and improvements in recent years are increasingly discussed.

Important characteristics to consider when designing a filter are computational complexity and efficiency, estimation accuracy, and robustness. A comparative overview of these characteristics for KF, EKF, UKF, and PF is given in Table 3.

Table 3. Overview of different KF based filtration methods characteristics.

	KF	EKF	UKF	PF
Assumption	Linear transition and observation Gaussian noise	Nonlinear transition and observation (Locally linear) Gaussian noise	Highly non-linear transition and observation Gaussian noise	Any arbitrary distribution Not only Gaussian noise
Estimation Accuaracy	Excat estimation accuaracy Satisfying all the assumptions give optimal output	Approximate (fixed by model - dependent on data) Satisfactory estimates	Approximate (fixed by model - dependent on data) Satisfactory estimates	Approximate (accuracy increrased with more particles) Smoother estimation
Computational Complexity	Low computational complexity	Propagation is sometimes complex and time-consuming Higher computational cost if the Jacobians can be computed numerically	Faster for implemetation Medium computational cost	Expensive asymptotic time complexity High memory request
Disadvantages	Defined only for linear systems	Only for systems with differentiable model Not optimal for highly nonlinear systems Diffucult to calculate Jacobians analytically	Not robust to a non-Gaussian environment	Insufficient number of particles lead to the failure of the algorithm

The combination of PF and KF can be called the particle KF. In doing so, PF is used for noise filtering, and KF is used to observe unmeasurable states. Changing the window width and the number of samples can affect the performance of such a filter. In another work, Jaros et al. (2018) [58] provided a comparative analysis of two particle KF formulations based on a kinematic model and/or a dynamic ship model. Similarly, Chen et al. (2018) [52] in the proposed algorithm used PF as a framework and additionally UKF as a density function for particle state optimization. This approach has proven satisfactory in terms of estimating the low-frequency oscillation of the ship.

In general, KF-based techniques demonstrate the ability to evaluate progressively the states using multi-sensor fusion, take the forward speed into account, are very effective in real-time estimations, and can evaluate measurements even when they are not available. On the other hand, the problem of smoothing, computational cost, and not sufficiently good estimation for the cases of individual performances depending on the data create disadvantages with these techniques and need to be further investigated. Research into the best combination of the described Kalman filter variations remains open, and it would be advisable to emphasize the possibility of finding an innovative solution that can be based on the complementarity of filtering algorithms (for noise filtering) and artificial intelligence algorithms (for predicting upcoming conditions).

7. Conclusions

An overview of various techniques for improving the performance of dynamic ship positioning systems was given. DP systems are very complex control techniques and are considered to be the most important part of the ship's equipment. To understand fully the need for such techniques in different sea conditions, as well as the effects of the environmental forces and moments, ship motions in all six degrees of freedom (with corresponding accelerations) are defined. Further, the system model is represented with joined filtration and processing algorithms. The mathematical modeling of DP systems principally depends on the expected conditions. The depth of the water, weather conditions, dimensions, and maximum ship speed determine such systems. It is important to filter out noisy measurements that are further used for control. Filtration can be done by utilizing the TF distributions, as well as their advanced, adaptive, data-driven, modified (high resolution) versions. To achieve more energy-efficiency, as well as safe and green ship routing, DP systems require information about DWS. New estimation techniques based on measured ship responses (precisely pitch, heave, and roll) and existing ship sensor devices are described with the analogy of wave height and direction measurements by moored buoys as conventional devices. The combination of motion responses for different encounter angles at certain sea conditions, together with the known transfers function, may improve the estimation accuracy, as presented. This approach usually utilizes KF, both for linear and nonlinear systems, and provides the ability to predict unknown information by combining multiple sensor measurements. Recent research on DP systems is focused on additional improvements of existing systems, taking into account all ship devices, as well as proper control, estimation and filtering algorithms. In future work, the authors aim to achieve better results in terms of improving the accuracy of processing and control algorithms in dynamic positioning systems. Based on the adaptation of the presented TF methods, the goal will be to obtain more appropriate extraction of the input characteristics of the ship's responses. Furthermore, using multiple ship responses at the same time, with an estimation of the wave amplitude, the estimation algorithm should be able to estimate the relative wave direction. A step further in improving the presented estimation techniques could also be achieved by an integration algorithm that uses one of the custom nonlinear versions of the Kalman filter in combination with a proper neural network. This way, the application of custom artificial intelligence algorithms to train neural networks and predict input signals may provide an innovative solution in later work.

Author Contributions: conceptualization, D.S., J.L., and J.P.-O.; methodology, D.S. and J.L.; software, D.S. and J.L.; validation, J.L. and J.P.-O.; formal analysis, D.S.; investigation, D.S. and J.P.-O.; resources, J.P.-O. and S.K.; data curation, J.P.-O. and S.K.; writing, original draft preparation, D.S.; writing, review and editing, J.L., J.P.-O. and S.K.; visualization, D.S.; supervision, J.L. and J.P.-O.; project administration, J.L. and J.P.-O.; funding acquisition, J.L. and J.P.-O. All authors read and agreed to the published version of the manuscript.

Funding: This work was fully supported by the Croatian Science Foundation under the projects IP-2018-01-3739 and IP-2020-02-4358, Center for Artificial Intelligence and Cybersecurity-University of Rijeka, University of Rijeka under the projects uniri-tehnic-18-17 and uniri-tehnic-18-15, and European Cooperation in Science and Technology (COST) under the project CA17137.

Conflicts of Interest: The authors declare no conflict of interest.

Abbreviations

The following abbreviations are used in this manuscript:

CWT	Continuous wavelet transform
DP	Dynamic positioning
DWS	Directional wave spectra
EKF	Extended Kalman filter
GNSS	Global Navigation Satellite Systems
GPS	Global positioning system
INS	Inertial navigation system
KF	Kalman filter
PF	Particle filter
RAO	Response amplitude operator
SM	S-method
SPWVD	Smoothed pseudo Wigner–Ville distribution
STFT	Short-time Fourier transform
TF	Time-frequency
UKF	Unscented Kalman filter
WVD	Wigner–Ville distribution

References

1. Nielsen, U.D. A concise account of techniques available for shipboard sea state estimation. *Ocean Eng.* **2017**, *129*, 352–362. [CrossRef]
2. Pascoal, R.; Perera, L.; Soares, C.G. Estimation of directional sea spectra from ship motions in sea trials. *Ocean Eng.* **2017**, *132*, 126–137. [CrossRef]
3. Nielsen, U.D.; Brodtkorb, A.H.; Sørensen, A.J. Sea state estimation using multiple ships simultaneously as sailing wave buoys. *Appl. Ocean Res.* **2019**, *83*, 65–76. [CrossRef]
4. Pikula, S.; Beneš, P. Comparison of measures of time-frequency distribution optimization. In Proceedings of the 8th International Congress on Ultra Modern Telecommunications and Control Systems and Workshops (ICUMT 2016), Lisbon, Portugal, 18–20 October 2016; pp. 314–319. [CrossRef]
5. Sandsten, M. Lunds Universitet-Matematikcentrum. Available online: http://www.maths.lu.se/fileadmin/maths/personal_staff/mariasandsten/TFkompver4.pdf (accessed on 15 March 2020).
6. Yang, Y.; Peng, Z.; Zhang, W.; Meng, G. Parameterized time-frequency analysis methods and their engineering applications: A review of recent advances. *Mech. Syst. Signal Process.* **2019**, *119*, 182–221. [CrossRef]
7. Fu, B.; Liu, J.; Wang, Q. Multi-Sensor Integrated Navigation System for Ships Based on Adaptive Kalman Filter. In Proceedings of the IEEE International Conference on Mechatronics and Automation (ICMA 2019), Tianjin, China, 4–7 August 2019; pp. 186–191. [CrossRef]
8. Vaernø, S.A.; Skjetne, R.; Kjerstad, Ø.K.; Calabrò, V. Comparison of control design models and observers for dynamic positioning of surface vessels. *Control Eng. Pract.* **2019**, *85*, 235–245. [CrossRef]
9. Peng, X.; Zhang, B.; Rong, L. A robust unscented Kalman filter and its application in estimating dynamic positioning ship motion states. *J. Mar. Sci. Technol.* **2019**, *24*, 1265–1279. [CrossRef]
10. Liu, R.; Yang, X.; Jia, Y. Design of nonlinear observer for ship dynamic positioning system. In Proceedings of the IEEE 3rd Information Technology, Networking, Electronic and Automation Control Conference (ITNEC 2019), Chengdu, China, 15–17 March 2019; pp. 1339–1342. [CrossRef]
11. Ding, L. Kalman Filtering Applied to Wave Spectrum Estimation Based on Measured Vessel Responses. Master's Thesis, Technical University of Denmark, Lyngby, Denmark, 2016.
12. Sarwito, S.; Semin; Zaman, M.B.; Soedibyo; Ramadhan, T.S. Short circuit study on closed circuit electrical system of ship dynamic positioning system based on laboratory scale experiment. *AIP Conf. Proc.* **2019**, *2187*, 060004. [CrossRef]
13. Shigunov, V.; Moctar, O.E.; Rathje, H. Operational Guidance for Prevention of Cargo Loss and Damage on Container Ships. *Ship Technol. Res.* **2010**, *57*, 8–25. [CrossRef]
14. Fossen, T.; Perez, T. Kalman filtering for positioning and heading control of ships and offshore rigs. *IEEE Control Syst. Mag.* **2009**, *29*, 32–46. [CrossRef]
15. Fossen, T.I. *Handbook of Marine Craft Hydrodynamics and Motion Control*; John Wiley & Sons, Ltd: Trondheim, Norway, 2011; ISBN 978-1-119-99149-6.
16. Tomera, M. Dynamic Positioning System for a Ship on Harbour Manoeuvring with Different Observers. Experimental Results. *Pol. Marit. Res.* **2014**, *21*, 13–24. [CrossRef]
17. Park, K.P.; Jo, A.R.; Choi, J.W. A study on the key performance indicator of the dynamic positioning system. *Int. J. Nav. Archit. Ocean Eng.* **2016**, *8*, 511–518. [CrossRef]
18. Smierzchalski, R. The structure of the control system for a dynamically positioned ship. In Proceedings of the 21st International Conference on Methods and Models in Automation and Robotics (MMAR 2016), Miedzyzdroje, Poland, 29 August–1 September 2016; pp. 641–644. [CrossRef]
19. Tannuri, E.; Agostinho, A.; Morishita, H.; Moratelli Junior, L. Dynamic positioning systems: An experimental analysis of sliding mode control. *Control Eng. Pract.* **2010**, *18*, 1121–1132. [CrossRef]
20. Tomera, M. Dynamic positioning system design for "Blue Lady". Simulation tests. *Pol. Marit. Res.* **2012**, *19*, 57–65. [CrossRef]
21. Zalewski, P. GNSS Integrity Concepts for Maritime Users. In Proceedings of the European Navigation Conference (ENC 2019), Warsaw, Poland, 9–12 April 2019; pp. 1–10. [CrossRef]
22. Malik, Z.M. Design and Implementation of Temporal Filtering and Other Data Fusion Algorithms to Enhance the Accuracy of a Real Time Radio Location Tracking System. Master's Thesis, Department of Electronics, Mathematics and Natural Sciences, University of Gävle, Gävle, Sweden, 2012.

23. I. Fossen, T.; Strand, J.P. Nonlinear Passive Weather Optimal Positioning Control (WOPC) System for Ships and Rigs: Experimental Results. *Automatica* **2001**, *37*, 701–715. [CrossRef]

24. Milne, I.; Zed, M. Deriving directional wave spectra from ship motions. In Proceedings of the Offshore Technology Conference Asia (2018), Kuala Lumpur, Malaysia, 20–23 March 2018; Volume 2, pp. 876–886.

25. Nielsen, U.; Galeazzi, R.; Brodtkorb, A. Evaluation of shipboard wave estimation techniques through model-scale experiments. In Proceedings of the OCEANS (2016), Shanghai, China, 10–13 April 2016; pp. 1–8. [CrossRef]

26. Nielsen, U.D. Estimation of Directional Wave Spectra from Measured Ship Responses. Ph.D. Thesis, Technical University of Denmark, Lyngby, Denmark, 2005.

27. Nielsen, U. Estimations of on-site directional wave spectra from measured ship responses. *Mar. Struct.* **2006**, *19*, 33–69. [CrossRef]

28. Bjerregard, M. Methods for Sea State Estimation. Master's Thesis, Technical University of Denmark, Lyngby, Denmark, 2014.

29. Nielsen, U.; Iseki, T. A Study on Parametric Wave Estimation Based on Measured Ship Motions. *J. Jpn. Inst. Navig.* **2012**, *126*, 171–177. [CrossRef]

30. Nielsen, U.; Andersen, I.; Koning, J. Comparisons of Means for Estimating Sea States from an Advancing Large Container Ship. In Proceedings of the 12th International Symposium on PRActical Design of Ships and Other Floating Structures (PRADS 2013), CECO, Changwon City, Korea, 20–25 October 2013.

31. Montazeri, N.; Nielsen, U.; Jensen, J. Estimation of wind sea and swell using shipboard measurements—A refined parametric modeling approach. *Appl. Ocean Res.* **2016**, *54*, 73–86. [CrossRef]

32. Iseki, T. Real-time Estimation of Directional Wave Spectra Using Non-Stationary Ship Motion Data. *J. Jpn. Inst. Navig.* **2009**, *120*, 117–123. [CrossRef]

33. Pascoal, R.; Guedes Soares, C. Non-parametric wave spectral estimation using vessel motions. *Appl. Ocean Res.* **2008**, *30*, 46–53. [CrossRef]

34. Pascoal, R.; Guedes Soares, C. Kalman filtering of vessel motions for ocean wave directional spectrum estimation. *Ocean Eng.* **2009**, *36*, 477–488. [CrossRef]

35. Nielsen, U.D. Deriving the absolute wave spectrum from an encountered distribution of wave energy spectral densities. *Ocean Eng.* **2018**, *165*, 194–208. [CrossRef]

36. Nielsen, U.D.; Brodtkorb, A.H.; Sørensen, A.J. A brute-force spectral approach for wave estimation using measured vessel motions. *Mar. Struct.* **2018**, *60*, 101–121. [CrossRef]

37. de Souza, F.; Tannuri, E.; de Mello, P.; Franzini, G.; Mas-Soler, J.; Simos, A. Bayesian estimation of directional wave-spectrum using vessel motions and wave-probes: Proposal and preliminary experimental validation. *J. Offshore Mech. Arct. Eng.* **2018**, *140*. [CrossRef]

38. Boashash, B. Chapter 6—Advanced Implementation and Realization of TFDs. In *Time-Frequency Signal Analysis and Processing*, 2nd ed.; Boashash, B., Ed.; Academic Press: Oxford, UK, 2016; pp. 331–385. [CrossRef]

39. Shafi, I.; Jamil, A.; Shah, S.; Kashif, F. Techniques to Obtain Good Resolution and Concentrated Time-Frequency Distributions: A Review. *EURASIP J. Adv. Signal Process.* **2009**, *2009*, 43. [CrossRef]

40. Feng, Z.; Liang, M.; Chu, F. Recent advances in time–frequency analysis methods for machinery fault diagnosis: A review with application examples. *Mech. Syst. Signal Process.* **2013**, *38*, 165–205. [CrossRef]

41. Auger, F.; Flandrin, P.; Gonçalves, P.; Lemoine, O. *Time-Frequency Toolbox Tutorial*; CNRS: Paris, France; Rice U.: Houston, TX, USA, 2005.

42. Sejdic, E.; Djurovic, I.; Jiang, J. Time–frequency feature representation using energy concentration: An overview of recent advances. *Digit. Signal Process.* **2009**, *19*, 153–183. [CrossRef]

43. Lerga, J.; Sucic, V. Nonlinear IF Estimation Based on the Pseudo WVD Adapted Using the Improved Sliding Pairwise ICI Rule. *Signal Process. Lett. IEEE* **2009**, *16*, 953–956. [CrossRef]

44. Torsvik, T.; Soomere, T.; Didenkulova, I.; Sheremet, A. Identification of ship wake structures by a time–frequency method. *J. Fluid Mech.* **2015**, *765*, 229–251. [CrossRef]

45. Pethiyagoda, R.; McCue, S.; J. Moroney, T. Spectrograms of ship wakes: Identifying linear and nonlinear wave signals. *J. Fluid Mech.* **2016**, *811*, 189–209. [CrossRef]

46. Didenkulova, I.; Sheremet, A.; Torsvik, T.; Soomere, T. Characteristic properties of different vessel wake signals. *J. Coast. Res.* **2013**, *65*, 213–218. [CrossRef]

47. Sheremet, A.; Gravois, U.; Tian, M. Boat-Wake Statistics at Jensen Beach, Florida. *J. Waterw. Port Coast. Ocean Eng.* **2013**, *139*, 286–294. [CrossRef]

48. Pethiyagoda, R.; Moroney, T.J.; Macfarlane, G.J.; Binns, J.R.; McCue, S.W. Time-frequency analysis of ship wave patterns in shallow water: Modelling and experiments. *Ocean Eng.* **2018**, *158*, 123–131. [CrossRef]

49. Torsvik, T.; Herrmann, H.; Didenkulova, I.; Rodin, A. Analysis of ship wake transformation in the coastal zone using time–frequency methods. *Proc. Est. Acad. Sci.* **2015**, *64*, 379–388. [CrossRef]

50. Yu, D.; Wang, Y. Time-frequency feature analysis of naval vessel impact response using the Hilbert Huang transform method. *J. Mar. Eng. Technol.* **2013**, *12*, 35–45. [CrossRef]

51. Orderud, F. Comparison of Kalman Filter Estimation Approaches for State Space Models with Nonlinear Measurements. Available online: http://citeseerx.ist.psu.edu/viewdoc/summary?doi=10.1.1.135.9250 (accessed on 15 March 2020).

52. Chen, Y.; Yang, X.; Liu, R. A Nonlinear Sate Estimate for Dynamic Positioning Based on Improved Particle Filter. In Proceedings of the 2nd IEEE Advanced Information Management, Communicates, Electronic and Automation Control Conference (IMCEC 2018), Xi'an, China, 25–27 May 2018; pp. 880–884. [CrossRef]

53. Alminde, L.; Bendtsen, J.D.; Stoustrup, J. A quantized state systems approach for Jacobian free Extended Kalman filtering. *IFAC Proc. Vol.* **2007**, *40*, 469–474. [CrossRef]

54. Küchler, S.; Eberharter, J.; Langer, K.; Schneider, K.; Sawodny, O. Heave Motion Estimation of a Vessel Using Acceleration Measurements. *IFAC Proc. Vol.* **2011**, *44*, 14742–14747. [CrossRef]

55. Assaf, M.; Petriu, E.; Groza, V. Ship track estimation using GPS data and Kalman Filter. In Proceedings of the IEEE International Instrumentation and Measurement Technology Conference: Discovering New Horizons in Instrumentation and Measurement (I2MTC 2018), Huston, TX, USA, 4–17 May 2018; pp. 1–6. [CrossRef]

56. Sørensen, A.J. A survey of dynamic positioning control systems. *Annu. Rev. Control* **2011**, *35*, 123–136. [CrossRef]

57. Wang, X.; Li, T.; Sun, S.; Corchado Rodríguez, J. A Survey of Recent Advances in Particle Filters and Remaining Challenges for Multitarget Tracking. *Sensors* **2017**, *17*, 2707. [CrossRef]

58. Jaroś, K.; Witkowska, A.; Śmierzchalski, R. Designing particle kalman filter for dynamic positioning. *Adv. Intell. Syst. Comput.* **2018**, *635*, 157–168, doi:10.1007/978-3-319-64474-5_13. [CrossRef]

Journal of
*Marine Science
and Engineering*

Article

Optimized Dislocation of Mobile Sensor Networks on Large Marine Environments Using Voronoi Partitions

Mario D'Acunto [1,*], Davide Moroni [2], Alessandro Puntoni [1] and Ovidio Salvetti [2]

[1] Institute of Biophysics, National Research Council of Italy, IBF-CNR, via Moruzzi 1, 56124 Pisa, Italy;
    alessandro.puntoni@cnr.it
[2] Institute of Information Science and Technologies, National Research Council of Italy, ISTI-CNR,
    Via Moruzzi 1, 56124 Pisa, Italy; davide.moroni@cnr.it (D.M.); ovidio.salvetti@isti.cnr.it (O.S.)
* Correspondence: mario.dacunto@cnr.it; Tel.: +39-05-0621-2763

Received: 6 January 2020; Accepted: 12 February 2020; Published: 18 February 2020

Abstract: The real-time environmental surveillance of large areas requires the ability to dislocate sensor networks. Generally, the probability of the occurrence of a pollution event depends on the burden of possible sources operating in the areas to be monitored. This implies a challenge for devising optimal real-time dislocation of wireless sensor networks. This challenge involves both hardware solutions and algorithms optimizing the displacements of mobile sensor networks in large areas with a vast number of sources of pollutant factors based mainly on diffusion mechanisms. In this paper, we present theoretical and simulated results inherent to a Voronoi partition approach for the optimized dislocation of a set of mobile wireless sensors with circular (radial) sensing power on large areas. The optimal deployment was found to be a variation of the generalized centroidal Voronoi configuration, where the Voronoi configuration is event-driven, and the centroid set of the corresponding generalized Voronoi cells changes as a function of the pollution event. The initial localization of the pollution events is simulated with a Poisson distribution. Our results could improve the possibility of reducing the costs for real-time surveillance of large areas, and other environmental monitoring when wireless sensor networks are involved.

Keywords: Voronoi partition; mobile sensor networks; wireless sensor networks; environmental monitoring; marine environment; oil spills

1. Introduction

A growing number of applications, such as spatial distribution mapping, dynamic sensors coverage, and environmental extensive area monitoring, have motivated the development of both sensing hardware and algorithms for target-oriented mobile sensor networks [1,2]. Any such application requires ad hoc customization of hardware and software solutions. Large area monitoring, for example, presents the main problem for biasing certain regions of interest. Indeed, some parts might be prioritized based on prior knowledge and due to an inherent time-constraint that prohibits an exhaustive search of the area; for instance, in the case of emergencies and search and rescue operations. This situation is met during the oil spill monitoring of large marine regions, and in general, in all pollution events involving diffusion [3]. The most probable scenario is that areas in which human beings, pollution activities, or hazardous materials are likely to be present should be searched first, while regions with less probability of containing these features are searched later. Consequently, both inferential statistical methods and mobile sensor networks able to navigate from an initial configuration within a region containing shaped obstacles were developed [4]. In this paper, we focus the attention on coordination algorithms to improve the capability of mobile sensors networks in covering and monitoring relevant portions of larger areas. Our proposed solution to this problem is obtained using centroidal Voronoi partitions.

The centroidal Voronoi partition is a widely used scheme of portioning a given space, and finds applications in many fields, such as image processing, sensor coverage, crystallography, and CAD [5–8]. The basic components of a Voronoi partition are:

(i)   A space, which is to be partitioned;

(ii)   A set of sites, or nodes or generators;

(iii)   A distance measure, such as the Euclidean distance.

In the case of homogeneous sensors with a typical circular sensing area, the sensor located in a Voronoi cell V_i is closest to all the points $q \in V_i$, and hence, by the strictly decreasing variation of sensors effectiveness with distance, the sensor is most effective within V_i, provided the circular area covered by the sensor does not cover the entire cell. Thus, the Voronoi decomposition leads to optimal partitioning of the space, in the sense that each sensor is most effective within its corresponding Voronoi cell. In the heterogeneous case too, it is easy to see that each sensor is most effective in its own generalized Voronoi cell. Since the partitioning is optimal, it is necessary to find the location of each sensor within its generalized Voronoi cell. Voronoi-based approaches have been used in the recent past for optimizing the coverage of mobile sensor networks. For instance, in [9], the Maxmin-vertex and Maxmin-edge algorithms are proposed to maximize the minimum distance of every sensor form the vertices and edges respectively of its Voronoi cells. In [10], such results are improved by also taking into consideration the adaptive ranges of sensors based on their respective residual energy. In [11], the authors proposed to consider a heterogeneous network made of both static and mobile sensors, and a bidding protocol for the placement of the mobile ones is proposed. Intuitively, mobile sensors are treated as servers to heal possible coverage holes left by the network as a whole. In this framework, optimization is also achieved through the distributed calculation of the Voronoi partition. With respect to these previous works that have shown the powerfulness of the Voronoi-based approaches in coverage improvement, in this paper, a further element is considered which is of relevance when dealing with environmental monitoring; i.e., adaptive prioritizing of areas to be monitored.

In the specific case of marine monitoring, adaptability is necessary to deal with (i) changing conditions in maritime traffic, weather and currents, and (ii) available knowledge of pollution events that have already taken place. This variable information should be put in relation to the variable impacts that events can have on different areas; e.g., due to the existence of marine parks, the presence of endangered animal species and proximity to shores.

Indeed, spills of oil and related petroleum products in marine environments can have a severe biological and economic impact [12,13]. With modern remote sensing instrumentation, such pollution events can be monitored on the open ocean around the clock. The monitoring of large marine areas generally involves complex marine information systems [14], which act as catalysts to integrate data from multiple and disparate data sources. Each source provides a specific piece of information, and through a suitable fusion and correlation process, they offer as a whole, a comprehensive characterization of the status of a marine area in terms of pollution events, maritime traffic, weather, and oceanic currents. In this process, key sources are SAR and optical satellite imaging to discover and quantify both oil slicks and real vessel traffic; the Automatic Identification System (AIS) for real-time identification and trajectory analysis of vessels passing through an area; and airborne hyperspectral sensors for detection and precise characterization of oil spills.

Sensorized buoys and surface and underwater vehicles can also be used to enforce remote monitoring. Indeed, autonomous underwater vehicles (AUVs) and mobile buoys equipped with different sensor devices are excellent candidates for real-time monitoring of oil spills through areas where the pollution events can take place. A new generation of buoys dedicated to oil spill monitoring is becoming available in a prototypal shape [15] or in their commercial evolution. Other interesting sources of information are based on crowd-sensing methodologies [16].

Environmental decision support systems (EDSS) are employed to orchestrate and make optimal and sustainable use of available monitoring and interventional resources. According to the survey [17], routine goals are (i) to actuate policies for confirming partial observations obtained by one source by collecting further information and (ii) to prioritize monitoring according to an adaptive risk map. For instance, if a slick is detected by remote sensing, in situ sensors such as buoys or AUVs can be deployed to assess its presence. Additionally, marine currents and weather conditions can increase the impact and the area affected by the possible pollution events.

The results presented in this paper were obtained having in mind (but are not limited to) the monitoring of large marine areas where oil spill events can take place, and where the diffusion evolution of any oil slick is known a priori. In addition, we consider marine sensors equipped with sensing hardware to consist mainly of an electronic nose, a bathymetric sensor, a GPS device, an anemometer, a motor for self-dislocation depending on a pollution event, and a complete communication set up with a remote station.

Actually, our simulation considers minor influences of weather and low marine waves, and low velocity for pollution event drift, so that the diffusion dynamics can be considered slow with respect to the sensor network's self-dislocation, which can be considered fast. The pollution events have been considered uncorrelated and were simulated with a Poisson distribution. Similar results were obtained with non-Poisson distributions.

The results show the capacity and the limits of event-dependent real-time monitoring of a sensor network. One of the limiting aspects is due to the real hardware facilities, especially for what regards real-time communication between the sensors and a remote station. The communication set up as considered in this paper is rather idealized; nevertheless, it is in line with the recent communication capabilities that will be available on sensorized devices for real-time monitoring of large marine areas.

The paper is organized as follows. In Section 2, we introduce the modeling of pollution events using a Poisson distribution and describe the Voronoi tessellation arising from a number of geolocalized events. In Section 3, details about Voronoi partitions, both classical and generalized, are given. In Section 4, we introduce an objective function to be maximized for optimal monitoring of an area subject to a static risk of pollution events. In this scenario, mobile sensors should follow a certain physical velocity field to move from their initial configurations to the optimal one. In Section 5, we assume that the risk of pollution changes over time in response to external events, diffusion and weathering; in this scenario, an iterative algorithm is proposed for letting the sensor network dislocate itself in an optimal way. Further, experimental simulations are described. Finally, in Section 6 we derive our conclusions.

The main contributions of the paper can be summarized as follows:

- A solution for the optimal displacement of a sensor network is proposed for the monitoring of large areas. The solution is based on formalized geometrical methods using Voronoi centroidal tessellation.
- The proposed solution is fitted into an iterative algorithm, which can evolve in real-time and continuously, the displacement of the network in response to dynamic events and changes in risk.
- The results are proven theoretically and with numerical simulations.

2. Modeling of Pollution Events

The first approach for monitoring a large area is given by the regular and symmetric dislocation of the sensors. Nevertheless, this approach is not practicable if the number of sensors with a finite radius of sensing power is not enough to cover the large area to be monitored. In addition, we have to consider the dislocation as event-dependent; i.e., considering the possible pollution source. An unwanted event is a random variable. Therefore, the immediate suggestion for a suitable event-dependent dislocation of a number of sensor networks is given by the Voronoi tessellations. A great number of natural phenomena are described by Voronoi tessellations; in this section, we try to respond to the main question: why should the optimal dislocations in a large area of a network composed by sensors with circular sensing power follow a centroidal Voronoi tessellation? To give an answer to this question, the starting point is to introduce the probability density function (PDF) for the occurrence of a pollution event, x. The reasonable shape of a PDF for a stochastic event is a Poisson distribution:

$$H(x) = \lambda \exp(-\lambda x)dx \tag{1}$$

where λ is the hazard rate of the exponential distribution. Let us consider the sum of two uncorrelated events, y, the variable change, y = 2x, and x = 1/λ. The PDF (Equation (1)) can be written as:

$$H(x) = 2x\exp(-2x)d(2x) \tag{2}$$

The result in Equation (2) can be written using a gamma variate distribution [18,19] as a function of the space dimension d; in our case d = 2.

$$H(x;d) = \frac{2d}{\Gamma(2d)}(2dx)^{2d-1}\exp(-2dx) \tag{3}$$

where, now, $0 \le x < \infty$, and $\Gamma(2d)$ is the gamma function with argument 2d. That PDF (Equation (3)) has a mean of $\mu = 1$ and variance $\sigma^2 = 1/d$. The gamma PDF can be generalized by introducing new additional parameters that take into account the complexity of the real situation, pollution sources, environmental weather, etc. In this paper, we restrict ourselves to gamma PDF as in Equation (3). Figure 1 shows the distribution as given by Equation (3) of pair events distance; i.e., the distance between two events close one each other.

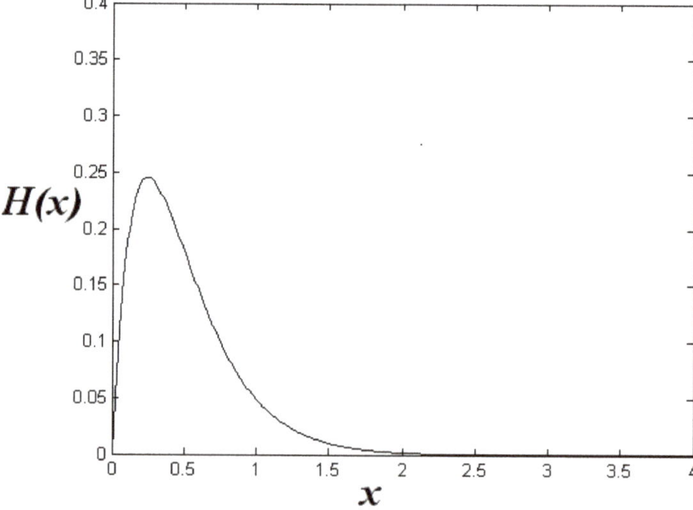

Figure 1. Distribution of pair events distance as given by a Poisson probability density function (PDF), Equation (3).

Let us consider n pollution events, distributed randomly in a square area. The monitoring of such diffusion processes can be made considering the equivalent Voronoi diagram, where the vertexes are coincident with the pollution events coordinates (see Figure 2). Similar results are obtained when the events are located using a Gaussian distribution. This is the case of pollution events with a correlation, anywhere the result does not present a significant difference.

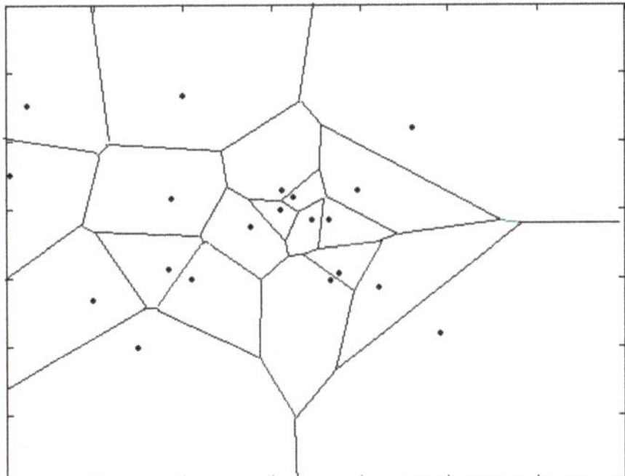

Figure 2. Example of dislocations of n = 20 pollution event randomly distributed using a Poisson PDF. The correspondent Voronoi diagram is built considering the pollution events located at the nodes.

3. Basic Properties for a Voronoi Partition

A standard Voronoi partition can be introduced as follows: A collection $\{W_i\}$, $i \in \{1,2, \ldots ,N\}$ of subsets of a space X with disjoint interiors is said to be a partition of X if $\cup_i W_i = X$. Let $Q \subset \mathfrak{R}^d$ be a convex polytope in d-dimensional Euclidean space. Let $P = \{p_1, p_2, \ldots , p_N\}$, $p_i \in Q$ be the set of nodes of generators in Q. The Voronoi partition generated by P with respect to the Euclidean norm is the collection $\{V_i(P)\}$, $i \in \{1, 2, \ldots , N\}$, and is defined as:

$$V_i(P) = \left\{ q \in Q \middle| \|q - p_i\| \leq \|q - p_j\|, \forall p_j \in P \right\} \tag{4}$$

where $\| \cdot \|$ denotes the Euclidean norm. The Voronoi cell V_i is the collection of those points that are closest (with respect to the Euclidean metric) to p_i compared to any other point in P. In $\mathfrak{R}^2$, the boundary of each bounded Voronoi cell is the union of a finite number of line segments forming a closed $^\circ C$ curve. For example, the intersection of any two Voronoi cells can be null, a line segment, or a point. In d-dimensional space, the boundaries of the Voronoi cells are unions of convex subsets of at most d-1 dimensional hyperplanes in $\mathfrak{R}^d$, and the intersection of two Voronoi cells can be a convex subset of a hyperplane or a null set. Each of the Voronoi cells is a topologically connected non-null set. Generalizations of the above Voronoi basic partition to suit specific applications can be found in the literature [12,13]. A possible generalization of the Voronoi partition can be introduced considering a space $Q \subset \mathfrak{R}^d$, sand et of points called nodes or generators $P = \{p_1, p_2, \ldots , p_N\}$, $p_i \in Q$, with $p_i \neq p_j$, whenever $i \neq j$, and monotonically decreasing analytic functions $f_i: \mathfrak{R} + \rightarrow \mathfrak{R}$, where f_i is called a node function for the i-th node. It is possible to define a collection $\{V_i\}$, $i \in \{1, 2, \ldots ,N\}$, with mutually disjoint interiors, such that $Q = \cup_i V_i$, where V_i is now defined as:

$$V_i(P) = \left\{ q \in Q \middle| f_i\big(\|p_i - q\|\big) \geq f_i\big(\|p_j - q\|\big), \forall j \neq i, i,j \in \{1,2,\ldots,N\} \right\} \tag{5}$$

We call $\{V_i\}$, $i \in \{1, 2, \ldots , N\}$ a generalized Voronoi partition of Q with nodes P and node functions f_i. In this application, it should be noted that V_i can be topologically non-connected and may contain Voronoi cells. In addition, it should be noted that $q \in V_i$ means that the i-th sensor is the most effective in sensing at point q. In a standard Voronoi partition used in a homogeneous case, the $\leq$ sign for distances ensures that i-th sensor is most effective in V_i. Finally, the condition that f_i is analytic implies that for every $i, j \in \{1, 2, \ldots , N\}$, f_i-f_j is analytic. By the properties of a real analytic function, it derives that the set of intersection points between any two-node functions is a set of measure zero. This ensures that the intersection of any two cells is a set of measure zero; that is, the boundary of a cell is a set is made up of the union of at most d-1 dimensional subsets of $\mathfrak{R}^d$; otherwise, the requirements that the cells should have mutually disjoint interiors may be violated. The analyticity of the node functions $\{f_i\}$ is a sufficient condition to avoid this possibility.

4. Optimization in the Localization of Event-Dependent Mobile Sensors Network

In this Section, a solution to optimal monitoring of an area by means of mobile sensors with limited radial sensing ranges is proposed. Such sensors can either be mounted on buoys with motion capabilities or full-fledged AUVs or unmanned surface vehicles (USVs). As stated in the introduction, it is assumed that sensors can navigate quite fast with respect to the evolution of a pollution event by diffusion, marine currents, and weathering.

Let us consider a large area to be monitored by N sensors; the cost of real-time monitoring is rather high. Unwanted events to be monitored trigger a sensor network to navigate from an initial configuration towards a specific region within the large area where the probability of encountering unwanted events is higher.

Let $Q \subset \mathcal{R}^d$ be a convex polytope, the space in which the sensors have to be deployed, and $V_i \subset Q$ be the generalized Voronoi cell corresponding to the i-th node; we introduce a continuous density distribution function $\varphi: Q \rightarrow [0,1]$, where the density $\varphi(q)$ is the probability of an event of interest occurring in $q \in Q$, and $P = \{p_1, p_2, \ldots, p_N\}$, $p_i \in Q$ is the configuration of N sensors [20].

It is well known that the generalized Voronoi decomposition splits the objective function into a sum of contributions from each generalized Voronoi cell. As a consequence, the optimization problem can be solved in a spatially distributed manner; i.e., each sensor solving the part of objective function corresponding to its cell using only local information can achieve the optimal configuration.

Besides, another essential characteristic of the sensor network, as considered in this paper, is that any sensor possesses a limited range of action; with the sensing capability that decreases as a logarithm of the distance from the centroid. This limitation must be taken in due consideration. Let R_i be the limit on the range of the sensors and $f(p_i, R_i)$ be a closed disk centered at p_i with a radius R_i. The i-th sensor has access to information only from points in the set $V_i \cap f(p_i, R_i)$.

Let us consider the following objective function to be maximized:

$$H(P) = \sum_i \int_{V_i \cap f(p_i, R_i)} f_i\big(\|q - p_i\|\big)\varphi(q)dQ \tag{6}$$

where $\| . \|$ is the Euclidean distance and $f_i = 0$ if $\|q\text{-}p_i\| > R_i$. It is possible to define the derivative of H (P) as follows.

$$\frac{\partial H(P)}{\partial p_i} = \sum_j \int_{V_i} \left[\frac{\partial f_j\big(\|q - p_i\|\big)}{\partial p_i}\right]\varphi(q)dQ \tag{7}$$

In addition, it is possible to introduce the critical points defined as the points mass and the centroid of the V_i cells.

$$\frac{\partial H(P)}{\partial p_i} = \sum_j \int_{V_j} \left[\frac{\partial f_j\big(\|q-p_i\|\big)}{\partial p_i}\right]\varphi(q)dQ = \sum_j \int_{V_j} \left[\frac{\partial f_j(r_i)}{\partial r_i^2}\big(p_i - q\big)\right]\varphi(q)dQ =$$
$$\sum_i M_{V_i}\big(\hat{p}_i - p_i\big) \tag{8}$$

where $\hat{p}_i$ is the set of centroids where the mobile sensors must be dislocated, defined later. The centroids set has the property $\hat{p}_i \in Q$ because Q is an invariant set for Equation (8), while on the contrary, it is not guaranteed that $\hat{p}_i \in V_i$. The dynamical properties of the mobile sensors network to be dislocated in an optimal configuration make it possible to write the basic control law as

$$\dot{p}_i = -k\big(p_i - \hat{p}_i\big) \tag{9}$$

where the single mobile sensor moves toward $\hat{p}_i$ for $k > 0$. Using the properties of the gradient with respect to p_i as in Equation (8), we can write

$$\frac{\partial H(P)}{\partial p_i} = \sum_i M_{V_i \cap f(p_i, R_i)} \left(\hat{P}_{i \cap f(p_i, R_i)} - P_i \right) \tag{10}$$

One main problem for the optimal dislocation of the mobile sensors is the convergence to the ideal centroid locations when a pollution event must be monitored. In this case, the uncovered area could generate more difficulties in producing a real convergence, but this problem can be dropped using effective k parameters in Equation (9). The problem of an effective convergence of a common radius sensor network that is unable to cover a large area is very close to that of the class of heterogeneous sensor networks [20]. In our case, the effective k parameter implies that the mobile sensors have higher dislocating speed with respect to the evolution velocity of a pollution event. The pollution event, for example, an oil spill, modifies its shape on the marine surface due to weather conditions: if the event time-scale, τ, is $\tau < <1/k$, then it is reasonable to expect a fast convergence to the optimal centroids. The control law, Equation (9), drives the sensors towards their optimal configurations that correspond to the arrangement that guarantees the maximum sensing power at the lowest cost. This is because the closed-loop system for the sensors networks is modeled using a first-order dynamical system that is globally asymptotically stable for an effective choice of k parameters; this is translated in the time property of H (P), which we supposed to be continuous.

$$\frac{dH(P)}{dt} = \sum_i \frac{\partial H(P)}{\partial p_i} \dot{p}_i = 2 \sum_i \alpha_i M_{V_i \cap f(p_i, R_i)} \left(\hat{P}_{i \cap f(p_i, R_i)} - P_i \right)^2 \tag{11}$$

The Equation (11) assures that the initial configurations of localized sensors converge to the optimal centroids $\hat{p}_i$.

One special problem is the ability of the sensors network to relocate themselves as a function of a pollution event (for example, an oil spill) whose shape changes in time due to diffusion and weathering. If the velocity of the sensor network is faster than the characteristic oil slick time diffusion, the sensing power of the sensors network is hugely enforced. Suppose a diffusion process evolves in the convex polytope space Q: $Q \in \mathcal{R}2$, where ρ (x, y): $Q \to \mathcal{R}+$ can be used to represent the pollutant concentration over Q, where the dynamic process is modeled by a partial differential Equation (PDE). The possibility to reorganize the sensors, relocating them to enforce their sensing power, is generally energy consuming; for this reason, the control law for the converge of the sensors toward their final centroids location must be constrained to the sensing power, which in this work we consider as $E_{sensing}$ $(r_i) = a \cdot r_i$, for $r_i < R_i$ and $E_{sensing}$ $(r_i) = 0$ otherwise, where a is constant and i is referred to the i-th sensor.

If the pollution event velocity in a given direction is v_{event}; then the control law (9) must be written as

$$\dot{P}_i = \begin{cases} -\alpha \frac{(p_i - \hat{p}_i)}{\|p_i - \hat{p}_i\|} & \text{if} \quad \|p_i - \hat{p}_i\| \geq \mu \\ -\alpha \left(p_i - \hat{p}_i \right)/\mu & \text{otherwise} \end{cases} \tag{12}$$

where $\alpha = v_{sensor}/v_{event}$ is a parameter denoting how much faster the mobile sensor must be with respect to pollution velocity v_{event}; μ is a positive defined value taking into account the distance from the centroids to which converge. The control law (Equation (12)) makes the sensors move toward their respective centroids with a constant speed v_{sensor} when they are at a distance further than μ from the corresponding centroids and slow down as they approach them. The control law (Equation (12)) is calculated at regular steps using new centroids $\hat{p}_i$ as a function of the event. Suppose that at time t = 0, a pollution event is located at x0, y0: the initial shape $\rho 0$ is a point. Under the effects of environmental parameters, ρ changes with time, and then at regular time steps, the centroids set is recalculated and the re-dislocation of mobile sensors occurs, consequently, until the highest sensing activity is made when two centroids set are coincident during two subsequent time steps checks.

Let us consider now, the possibility of a dislocation of the sensors network from an initial set of centroids to a new set of centroids when an unwanted event must be taken into consideration. When the time evolution of pollutant concentration ρ (x, y, t) is known, it is possible to calculate masses and centroids on region V_i at regular time steps as

$$M_{V_i} = \int_{V_i} \rho\big(|q - p_i|\big)dq \text{ and } \hat{p}_i = \frac{\int_{V_i} q\rho\big(|q - p_i|\big)dq}{\int_{V_i} \rho\big(|q - p_i|\big)dq} \tag{13}$$

The masses and centroids are recalculated using a sampling time, generally a fraction of α using an algorithm very close to Lloyd Centroidal algorithm [21–24] applied to sensors with a limited sensing radius, as in Figure 3. When possible, the dynamical changes of $\rho(|q\text{-}p_i|,t)$ are usually modeled by a partial differential equation [13]. Because in many real situations the shape of $\rho(|q\text{-}p_i|,t)$ is unknown, we advance the idea that $\rho(|q\text{-}p_i|,t)$ could be substituted by a Poisson distribution of random events, so that $\rho(|q\text{-}p_i|,t)$ can occur in one point, as described by Figure 2, and evolve with low dynamics. In this manner, the n sensors can dislocate themselves to develop higher sensing power, thereby neutralizing the pollution as quickly as possible without making the area of interest overdosed.

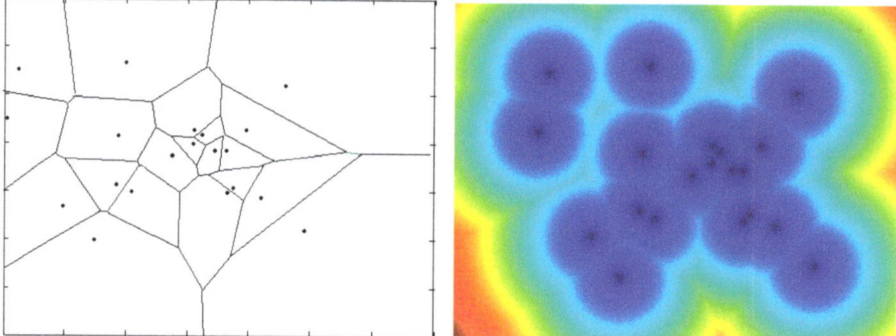

Figure 3. On the left, dislocations of n = 20 sensors, as in Figure 2; on the right, the representation of the power monitoring using the same dislocation configuration, where the sensors having common limited range R are located on the points defined by the Poisson distribution, Equation (3); the pollution event occupies one site. The blue domains represent the highest sensing power, and the black circles denoted the centroid set $\hat{p}_i$. Note that the area on the right image is slightly larger with respect to the left image.

5. Numerical Results and Discussion

Our method is based on the assumption that the occurrence of a random pollution event is an element of a more general class of events that follows a Poisson distribution, as described in Figures 1–3. The sensors' capability can be considered more effective if they can be located in a general configuration, where the pollution event occupies one site, and the sensors the other ones. Figure 3 represents both the Voronoi diagram and the corresponding power sensing using sensors with limited sensing radii. As the pollution event, described by $\rho(|q\text{-}p_i|,t)$, changes due to environmental conditions, the sensors must be re-dislocated on the positions coinciding with the centroids calculated using Equation (13). Because the shape of $\rho(|q\text{-}p_i|,t)$ is generally unknown, we can hypothesize that at regular time steps, the new location of $\rho(|q\text{-}p_i|,t)$ is on a position still described by the same Poisson distribution, that in our model represents a primitive partitions of space, which the Voronoi diagram represents the most effective operability of the sensors networks (Figure 3).

The new algorithm proposed for the control made by a mobile sensors network of a pollutant event occurring in a large environment can be described as follows:

(1)   Initial setting of the sensors $p_i \in \{p_1, \dots, p_n\}$; response time $t = 0$, coincident with a Poisson distribution of locations.

(2)   Computation of the Voronoi region V_i. All the sensors denoted by a standard limit range radius are defined as $R_i = A^{1/2}/(n + m)$, where $A = [0,1] \times [0,1]$ is the area to be monitored, *n* is the number of sensors, and *m* is an entire number taking into account the range of the uncontrolled area.

(3)   A pollution event takes place at time $t = 0$, located at a point as described by a Poisson distribution and the diffusion process described by $\rho(x, y, t)$; or equivalently $\rho(|q\text{-}p_i|,t)$, changes slowly.

(4)   Computation of centroid set $\hat{p}_i$ using Equation (13).

(5)   Mobile sensor networks move towards pollution location with a velocity described by α, considering the closer sensors located in the direction of less pollution shape variation using the control law (Equation (12)).

(6)   Iterative computations of new centroid sets $\hat{p}_{i-j}$ are made at regular times j, $j < i$. At any new estimate, the sensor nearest the slick is fixed and all the other sensors are moved following the new centroids set.

(7)   The optimal final configuration embedding the diffusion pollution event is reached.

The numerical results were obtained using a node function coinciding with power sensing as $f_i(r_i) = (R - r_i)^2$ for $r < R_i$, and $f_i(r_i) = 0$; otherwise, at any time step when the centroids set is recalculated, all the sensors are denoted by a common limit range radius defined as $R_i = A^{1/2}/(n + m)$, where *n* is the number of sensors and *m* is an entire number taking in account the uncontrolled area. For example, in Figure 3, we have considered n = 20 sensors distributed following the PDF (3), with m = 1. m being strictly connected to different hardware power sensing, running service, and environmental conditions, other than aging. It can be expressed by the empirical relation:

$$m \cong \frac{\omega \times \langle l^2_{vor} \rangle}{d_{diff}} \tag{14}$$

where ω is rate of possible pollution event due to ship traffic; for example, d_{diff} is a diffusion coefficient and describes the mobility of the sensor. $\langle l^2_{vor} \rangle$ denotes the averaged Voronoi cells' area.

The darker areas denote the areas monitored by the sensor network and the brighter regions the portion that is not effectively covered by the sensor monitoring. It is evident that while increasing the number of sensors, the area monitored increases, but this involves significant costs, reducing the benefits due to mobile sensors.

In many real applications, the area to be monitored is more extensive with respect to sensor network power sensing, m > >n. This leads to an initial dislocation that we can consider *blind*, because in our polytope space, $Q \subset \mathcal{R}^2$, we suppose the equiprobability for the occurrence of a pollution event. For example, in Figure 4 is shown a linear dislocation of seven sensors with a limited sensing radius (gray circles). We want to cover the maximum area with the lowest number of wireless sensors moving in a large area. The general characteristics of the mobility for the wireless sensors should meet this general requirement; nevertheless, large portions of the unchecked areas must be taken into consideration.

When a pollution event takes place, the sensors move in the direction of the pollutant slick, re-dislocating themselves in a Voronoi set centroid given by the knowledge of some environmental information. The environment changes the shape of the slick, expressed by $\rho(|q\text{-}p_i|,t)$, making it asymmetrical; consequently, the new dislocation implies that the new centroids overlap the individual sensing radii; i.e., $R_i - R_j < R$, with $i, j = 1 \dots n$, enforcing their power sensing in one direction rather than another one. The first step is a fast dislocation of a single sensor; for example, the sensor nearest to the pollution event. The control law (Equation (12)) is integrated calculating the centroids $\hat{p}_i$ at regular steps, once a pollution event has been located. The convergence is guaranteed by choosing an effective α parameter as ≥ 10 and time steps for the calculation of the new centroids set proportional to $\alpha/5$.

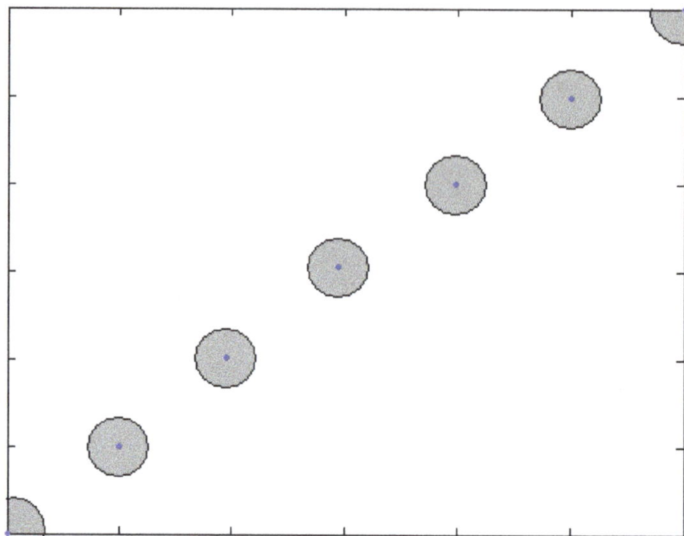

Figure 4. Dislocation of initial $n = 7$ sensors located in a blind linear configuration on a large area. The circle denotes the sensed area, m = 16n. A blind dislocation made by regular and equidistant positions provides equiprobability for the occurrence of a pollution event on the area to be monitored.

In Figure 5 the main results of our procedure are represented. The blind dislocation is changed as a function of pollutions events. In our case, we limited ourselves to a single pollution event moving but unchanging shape. The two localizations and the correspondent dislocations of sensor networks are displayed in Figure 5A and zoomed-in in Figure 5B. In a first approach, the blind dislocation is changed considering that the pollution event is localized at a point described by a Poisson distribution, where the pollution event and the other ones by the sensors occupy one site, following the Voronoi tessellation. The sensor nearest to the pollution event is locked and the other sensors are moved to be collocated at the positions identified by the centroids set calculated with the Equation (13). As the slick moves, the configuration of the sensor network is defined by the new centroids set, where at any new calculation, the nearest sensors are locked. The final result is reached when the sensing power is maximal. The sensing power is defined by the functions $f_i(r_i) = (R - r_i)^2$ for $r < R_i$, and $f_i(r_i) = 0$ otherwise. In Figure 5B, the final configuration is obtained by the five sensors forming a pentagon; the blue point denotes the sensors that in the preview centroids set were locked because of being nearest to the slick. The results in Figure 5 were obtained taking the parameter $\alpha = 10$; i.e., the sensor velocity has been taken to be ten times the slick velocity, a slow diffusion regime. This guarantees the fast converge to the centroids set without complicating the configuration due to the slick shift during the sensor dislocation. It is evident that the parameter α includes the main characteristics of the diffusion process. Fast dynamics of the slick, with fast changes of position and shape of the slick, would be useless without our algorithm.

All the numerical results were obtained by writing an original Matlab program. The presented results are rather general and they can be applied to many real situations when *a priori* knowledge of the environments to be monitored is unavailable. If *a priori* knowledge on where a pollution event can take place and the diffusion process, expressed by $\rho(|q-p_i|,t)$, is not available, the initial configuration of the sensors can be well simulated using the Poisson distribution of the random events for the centroids set. In such situations of real uncertainty, some improvements for our algorithm can be made conjugating the centroidal Voronoi tessellations with an inferential approach, but this is beyond the scope of the present paper.

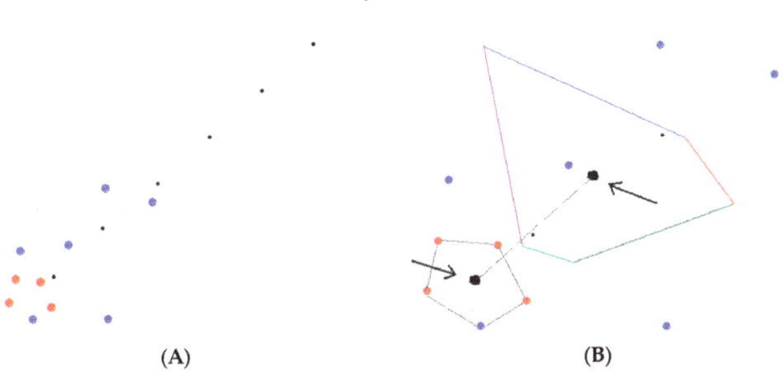

(A)   (B)

Figure 5. Evolution of the swarm composed by $n = 7$ sensors on large area starting from an initial set of blind positions as in Figure 4 and moving around a pollution events located at two different positions between the initial pollution occurrence and the final location when the sensors' power is maximal and real time monitoring is reached (initial and final black slick are indicated by the arrow in (**B**), and sensors are denoted by blue color in the intermediate configuration and red in the final configuration). The initial configuration changes are re-dislocating themselves in a new set of positions given by the centroids calculated with Equation (13) and using the control law Equation (12), $\alpha = 10$. (**B**) is a zoom-in of the (**A**).

6. Conclusions

The real-time surveillance of large areas requires the ability to dislocate sensor networks. Generally, the probability of the occurrence of a diffusion-based pollutant event depends on the burden of possible pollution sources and the spreading of pollution with time. This represents a challenge for devising real-time dislocation of the sensor network.

In this paper, we have presented theoretical and simulated results inherent to a centroidal Voronoi partition for the optimized dislocation of a set of heterogeneous buoys and similar sensor networks in a large area. The optimal deployment was found to be a variation of the generalized centroidal Voronoi configuration, where the sensors are located at the centroids of the corresponding generalized Voronoi cells. In turn, we provided a control law that under some constraints on the sensor velocity guarantees immediate results on stability and convergence of the centroids to the final position where the sensing power is higher. The generality of our findings could improve the possibility to apply our approach to reduce the costs of the real-time surveillance of large areas. The proposed solution has been fitted into an iterative algorithm that allows for evolution constantly, in real-time for the displacement of the network in response to dynamic events and changes in risk. Numerical simulations have been conducted to show the efficacy of the approach, while tests in a realistic scenario, consisting of areas of the Mediterranean Sea, are foreseen in the near future.

Author Contributions: M.D. conceived and developed the theoretical support of the methodology. M.D., D.M., A.P. and O.S. carried out the numerical results. All the authors discussed the results and contributed to the manuscript writing with a major contribution of M.D. and D.M. The overall study was supervised by M.D. All authors have read and agreed to the published version of the manuscript.

Funding: The work was partially supported by the Italian National Project PON "Smart Cities and Communities" S4E and by the EU FP7 Project Argomarine (Grant 234096).

Acknowledgments: The authors wish to thank the BIO-ICT joint laboratory between the Institute of Biophysics and the Institute of Information Science and Technologies, both of the National Research Council of Italy, in Pisa.

Conflicts of Interest: The authors declare no conflict of interest.

References

1. Ruiz-Garcia, L.; Lunadei, L.; Barreiro, P.; Robla, I. A review of wireless sensor technologies and applications in agriculture and food industry: State of the art and current trends. *Sensors* **2009**, *9*, 4728–4750. [CrossRef] [PubMed]

2. Dargie, W.; Poellabauer, C. *Fundamentals Of Wireless Sensor Networks: Theory And Practice*; John Wiley and Sons: Hoboken, NJ, USA, 2010.

3. Jarnelov, A. How to defend against future oil spills. *Nature* **2010**, *466*, 182–183. [CrossRef] [PubMed]

4. Li, Q.; Rus, D. Navigation protocols in sensor networks. *ACM Trans. on Sensors Networks* **2005**, *1*, 1–33. [CrossRef]

5. Cheddad, A.; Mohamad, D.; Abd Manaf, A. Exploiting Voronoi diagram properties in face segmentation and feature extraction. *Pattern Recognit.* **2008**, *41*, 3842–3859. [CrossRef]

6. Kalra, N.; Ferguson, D.; Stentz, A. Incremental reconstruction of generalized Voronoi diagrams on grids. *Robo. Auton. Syst.* **2009**, *57*, 123–128. [CrossRef]

7. Krishnamurthy, V.; Brostow, W.; Sochanski, J.S. Representation of properties of materials by Voronoi polyhedral. *Mater. Chem. Phys.* **1988**, *20*, 451–469. [CrossRef]

8. Papadopoulou, E. Critical area computation via Voronoi diagrams. *IEEE Trans. Comput. Aided Des. Integr. Circuits Syst.* **1999**, *4*, 463–474. [CrossRef]

9. Mahboubi, H.; Moezzi, K.; Aghdam, A.G.; Sayrafian-Pour, K.; Marbukh, V. Distributed deployment algorithms for improved coverage in a network of wireless mobile sensors. *IEEE Trans. Ind. Inform.* **2013**, *10*, 163–174. [CrossRef]

10. Mahboubi, H.; Aghdam, A.G.; Sayrafian-Pour, K. Toward autonomous mobile sensor networks technology. *IEEE Trans. Ind. Inform.* **2016**, *12*, 576–586. [CrossRef]

11. Wang, G.; Cao, G.; Berman, P.; La Porta, T.F. Bidding protocols for deploying mobile sensors. *IEEE Trans. Mob. Comput.* **2007**, *6*, 563–576. [CrossRef]

12. Eide, M.S.; Endresen, Ø.; Brett, P.O.; Ervik, J.L.; Røang, K. Intelligent ship traffic monitoring for oil spill prevention: Risk based decision support buildings on AIS. *Mar. Pollut. Bull.* **2007**, *54*, 145–148. [CrossRef] [PubMed]

13. Cocco, M.; Colantonio, S.; D'Acunto, M.; Martinelli, M.; Moroni, D.; Pieri, G.; Salvetti, O.; Tampucci, M. Information Technology Ocean Engineering. *CICC-ITOE Proc. Conf.* **2011**, *2*, 163–167.

14. Moroni, D.; Pieri, G.; Tampucci, M.; Salvetti, O. A proactive system for maritime environment monitoring. *Mar. Pollut. Bull.* **2016**, *102*, 316–322. [CrossRef] [PubMed]

15. Moroni, D.; Pieri, G.; Salvetti, O.; Tampucci, M.; Domenici, C.; Tonacci, A. Sensorized buoy for oil spill early detection. *Methods Oceanogr.* **2016**, *17*, 221–231. [CrossRef]

16. Martinelli, M.; Moroni, D. Volunteered geographic information for enhanced marine environment monitoring. *Appl. Sci.* **2018**, *8*, 1743. [CrossRef]

17. Moroni, D.; Pieri, G.; Tampucci, M. Environmental Decision Support Systems for Monitoring Small Scale Oil Spills: Existing Solutions, Best Practices and Current Challenges. *J. Mar. Sci. Eng.* **2019**, *7*, 19. [CrossRef]

18. Zaninetti, L. Poissonian and non-Poissonian Voronoi Diagrams with applications to the aggregations of molecules. *Phys. Lett. A* **2009**, *373*, 3223–3229. [CrossRef]

19. Zachary, D.S. The Inverse Poisson Functional for forecasting response time to environmental events and global climate change. *Sci. Rep.* **2018**, *8*, 1–9. [CrossRef]

20. Guruprasad, K.R.; Ghose, D. Automated Multi-Agent Search Using Centroidal Voronoi Configuration. *IEEE Trans. Autom. Sci. Eng.* **2011**, *8*, 420–423. [CrossRef]

21. Chao, H.; Chen, Y.Q.; Ren, W. Consensus of Information in Distributed Control of a Diffusion Process using Centroidal Voronoi Tessellations. In Proceedings of the 46th IEEE Conference on Decision and Control, New Orleans, LA, USA, 12–14 December 2007; pp. 1441–1446.

22. Chao, H.; Chen, Y.Q. Cooperative Sensing and Distributed Control of a Diffusion Process Using Centroidal Voronoi Tessellations. *Numer. Math. Theory Methods Appl.* **2010**, *3*, 162–177.

23. Cortes, J.; Martinez, S.; Karatas, T.; Bullo, F. Coverage control for mobile sensing networks. *IEEE Trans. Robot. Autom.* **2004**, *20*, 243–255. [CrossRef]

24. Du, Q.; Emeliarenko, M.; Ju, L. Convergence of the Lloyd algorithm for computing centroidal Voronoi tessellations. *SIAM J. Numer. Anal.* **2006**, *44*, 102–119. [CrossRef]

Journal of
Marine Science and Engineering

Article

A Layout Strategy for Distributed Barrage Jamming against Underwater Acoustic Sensor Networks

Mengchen Xiong [1], Jie Zhuo [1,*], Yangze Dong [2,*] and Xin Jing [1]

[1] School of Marine Science and Technology, Northwestern Polytechnical University, Xi'an 710072, China; xionngmc@mail.nwpu.edu.cn (M.X.); 589jingx@mail.nwpu.edu.cn (X.J.)
[2] National Key Laboratory of Science and Technology on Underwater Acoustic Antagonizing, Shanghai 201108, China
* Correspondence: jzhuo@nwpu.edu.cn (J.Z.); dongyangze@139.com (Y.D.)

Received: 9 March 2020; Accepted: 30 March 2020; Published: 3 April 2020

Abstract: Underwater acoustic sensor networks (UASNs) can effectively detect and track targets and therefore play an important role in underwater detection technology. To protect a target from being detected by UASNs, a distributed barrage jamming layout strategy is proposed, which considers the detection performance of UASNs as an indicator of the jamming performance. Since common indices of detection performance often involve specific signal processing methods, the Cramér–Rao bound (CRB) of multiple targets estimated by an UASN for distributed jammers is deduced in this paper, which is universal for all signal processing methods. The optimization model of the distributed jamming layout strategy is designed by maximizing the CRB to achieve the best jamming effect with limited jammers. A heuristic algorithm is used to solve this optimization model, and a numerical simulation shows that the optimal layout strategy for distributed jammers proposed in this paper achieves better performance than traditional jamming layout strategies. Considering the deviation of the position of the jammers from the ideal value due to the movement of water in a real marine environment, this paper also analyzes the jamming effects of strategies when there is error in the position of the jammers. The result proves the effectiveness and superiority of the proposed optimal layout strategy in an actual environment.

Keywords: distributed jamming technology; layout strategy; underwater acoustic sensor networks; Cramér–Rao bound

1. Introduction

In recent years, research on underwater acoustic sensor networks (UASNs) has received increasing attention. UASNs have great application prospects in the submarine field [1–4], including the monitoring of tsunamis and earthquakes and also military applications such as shore-based observation and target tracking. Compared with traditional detection systems, UASNs employ multiple sonar arrays working independently or collaboratively and can achieve direction measurements and also tracking and identification. Therefore, UASNs play an increasingly important role in the modern underwater detection field. With the development of related detection technologies, research on jamming strategies for UASNs is also attracting the attention of researchers [5]. The UASN jamming scenario usually includes several targets (marine vehicles such as ships and submarines) exposed in the UASN detection range. In order to avoid being detected by the UASN, the targets usually emit some noise jammers. Then the targets radiated signal received by the UASN is covered by jamming noise and the detection performance of the detector reduces. A well-designed jamming strategy can achieve the goal of protecting targets from being detected more effectively than other strategies and is an important research problem in underwater jamming technology.

Currently, the most common jamming method for underwater detection uses a single high-power jammer at a long distance. However, this approach cannot effectively jam all of the arrays in an UASN. A more flexible and effective jamming method should be applied in an underwater environment. Distributed jamming technology is a new type of jamming method [6] that uses many small low-cost jammers distributed near the receiving system that can jam multiple sensor arrays at the same time. Compared to traditional jamming methods, distributed jamming technology can achieve better results for UASNs.

Distributed jamming technology was first used to countermeasure radar systems. There are many studies on distributed jamming technology for radar systems with strong anti-jamming capabilities, such as multiple-input multiple-output (MIMO) radar [7–10] and networking radar [11]. However, most of these studies mainly focus on the power allocation of the jammers [7,8], without considering the influence of the locations of the distributed jammers. Due to the difference between the propagation medium and receiving equipment of the ground-based radio sensor networks and the underwater acoustic sensor networks [12], distributed jamming strategies in the radar field cannot be directly applied to UASNs. The receiver structures of UASNs are different from those of radar networks, so they have different detection methods, and the reflections of jamming effects against them are different. Related work in an underwater environment has only appeared in recent years. Jamming strategies for UASNs based on game theory were studied by Vadori and colleagues [13] and Xiao and co-workers [14]. In particular, the location of the jammer was studied by Vadori and colleagues [13], and the results showed that a jammer in some specific locations will have a stronger jamming effect on UASNs. However, research is limited to the case of a single jammer, and there is no description of the locations of distributed jammers.

In this paper, an optimal layout strategy for distributed jammers of UASNs is developed, which aims to minimize the target detection performance of the sensors by optimizing the location of the jammers, and thus protect the target from being attacked. Therefore, the detection performance of the UASNs can be applied as an evaluation index to evaluate the jamming performance. Since the same detection system using different signal processing methods will produce different estimation errors when estimating the target parameters, a more universal index should be considered. In a study by Zheng and colleagues [9], the Cramér–Rao bound (CRB) of MIMO radar was used to reflect the jamming performance, representing the minimum mean square error that the receiving system can achieve for an unbiased estimation of the target parameters. However, while the power allocation strategy of a single jammer was considered in their study [9], the distributed jammers and the effect of the jammers' location were not taken into consideration. In this paper, the CRB of an UASN is used to evaluate its detection accuracy. The larger the CRB, the larger the minimum estimation error that can be achieved by the receiver and the worse the parameter estimation performance of the receiver. Correspondingly, due to the effect of distributed jamming, the CRB of the receiver increases, which means that the jammers weaken the receiver's estimation performance for the target. Therefore, the CRB can be used as an evaluation index of the jamming effect. Based on this approach, an optimal distributed jamming layout strategy is proposed. By maximizing the CRB, the optimal locations of the distributed jammers can be obtained. In addition, since the calculation of the CRB is not limited to a specific signal processing method, the jamming layout strategy based on the CRB is applicable to a variety of receiving systems.

To obtain a CRB-based optimal model of the layout strategy for distributed jammers, the CRB of the target parameter estimation of UASNs under distributed jammers should be studied. There are studies on the CRB in receiver systems in the presence of multiple targets, mostly under the assumption of a uniform environmental noise field [15–18]. However, the jamming noise received by each node in the sensor network should be nonuniform and related under the influence of distributed jammers, and there is no relevant derivation in the existing literature. Therefore, in this paper, the CRB of UASNs in a nonuniform noise field under a distributed jamming strategy is derived, similar to the derivation used by Stoica and Nehorai [15] under the assumption of a uniform noise background.

According to the obtained CRB, a CRB maximization model with the position of the jammers as a variable is established and used as the basis of the distributed jamming optimization layout strategy. To solve this problem, particle swarm optimization (PSO) is adopted to analyze the distributed jamming strategy proposed in this paper through numerical experiments considering different environmental parameters. The proposed optimal layout strategy can cause higher CRB of the UASN than other traditional layout strategies, which means a stronger jamming effect to the UASN and stronger protection to targets.

The remainder of the paper is organized as follows. The system model of UASNs in the presence of multiple jamming sources is reviewed in Section 2. The CRB of UASNs for multiple target angle estimation in a jamming environment is derived in Section 3. The optimization model of the arrangement of distributed jammers is constructed in Section 4, and numerical results on the performance of the proposed jamming strategy are shown in Section 5. Conclusions are drawn in Section 6.

2. System Model

Before this section, some notational conventions used in this paper are shown in Table 1.

Table 1. Notations in this paper.

Notation	Description
$\mathrm{E}[\cdot]$	the expectation operator; for deterministic signals, $\mathrm{E}[\cdot]= lim_{N \to \infty}(1/N)\sum_{t=1}^{N}(\cdot)$
x^{H}	the conjugate transpose of x
x^{t}	the transpose of x
A	the matrix
$\delta_{m,n}$	the Dirac delta (=1 if m = n and 0 otherwise)

The jamming model studied in this paper consists of three parts: targets, jammers, and the UASN. Assuming they are in a two-dimensional plane, the geometric positions of the targets, jammers, and the UASN are shown in Figure 1. Underwater detection is usually achieved by sonar sensor array and uniform linear array (ULA) is the most common receiver. In this paper, the UASN is a sonar receiving system composed of multiple ULAs, which is also called a networking sonar system and widely used in underwater detection. Suppose the UASN contains L sensor arrays. Denote the center position of the lth receiving array as (x_{Rl}, y_{Rl}), the numbers of elements in the lth array as P_l, the space of each element in the lth array as Δ_l, and the tilt angle of the lth array relative to the x-axis as γ_l, where $l = 1, 2, \cdots, L$. Suppose there are M targets and K jammers. The position of the mth target is (x_{Tm}, y_{Tm}), and the position of the kth jammer is (x_{Jk}, y_{Jk}), where $m = 1, 2, \cdots, M$ and $k = 1, 2, \cdots, K$. The distances between the mth target and the kth jammer to the lth array are

$$d_{Rl}^{Tm} = \sqrt{(x_{Rl} - x_{Tm})^2 + (y_{Rl} - y_{Tm})^2} \tag{1}$$

$$d_{Rl}^{Jk} = \sqrt{\left(x_{Rl} - x_{Jk}\right)^2 + \left(y_{Rl} - y_{Jk}\right)^2} \tag{2}$$

The angle between the mth target and the lth array (relative to the array normal direction) is

$$\varphi_l^m = \arctan\left(\frac{x_{Rl} - x_{Tm}}{y_{Rl} - y_{Tm}}\right) + \gamma_l \tag{3}$$

Assume that the sensor arrays in Figure 1 are passive arrays and only receive radiation signals from the target and jammers and that there are a variety of environmental noise sources in the environment. Denote the radiation signals from the mth target as $s_m(t)$, the radiation signals from the kth jammer as $n_k(t)$, and the environmental noise influencing the elements of the lth array as $e_l(t)$. The jamming strategy involved in this paper is barrage jamming; that is, the target signal is covered by high-power noise. The research in this paper is based on the following assumptions:

Firstly: The jamming noise emitted by the interference source $n_k(t)$ is Gaussian white noise with an average value of 0 and a variance of σ_k. Each jammer transmits signals independently, which means that $En_k(t) = 0$, $En_k(t)n_k^H = \sigma_k I$, and $En_{k_1}(t)n_{k_2}^H = 0$. Assume that each element in the same array receives the same jamming power.

Secondly: All of the elements receive the same environmental noise power and are independent of each other. The environmental noise on the lth array $e_l(t)$ is Gaussian white noise with an average value of 0 and a variance of σ_0, where $l = 1, 2, \cdots, L$, $Ee_l(t) = 0$, and $Ee_l(t)e_l^H(t) = \sigma_0 I$.

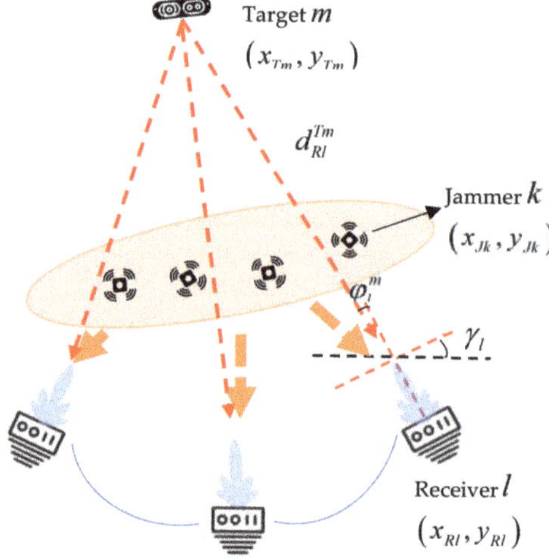

Figure 1. Model of the target, jammer, and sensor network.

Then the received signal model of the lth array can be expressed as

$$\mathbf{y_l}(t) = \mathbf{A_l}(\mathbf{\theta_T})\hat{\mathbf{s}}(t) + \mathbf{B_l}\hat{\mathbf{n}}(t) + e_l(t) \tag{4}$$

In Equation (4), $t = 1, 2, \cdots, N$. $\mathbf{y_l}(t) \in C^{P_l \times 1}$ is the noisy data vector. $\hat{\mathbf{s}}(t) \in C^{M \times 1}$ is the vector of the target radiation signal amplitudes, and $\hat{\mathbf{s}}(t) = [s_1(t), \cdots, s_m(t)]^t$. $\hat{\mathbf{n}}(t) \in C^{K \times 1}$ is the jamming noise, and $\hat{\mathbf{n}}(t) = [n_1(t), \cdots, n_K(t)]^t$. The matrix $\mathbf{A_l}(\mathbf{\theta_T}) \in C^{P_l \times M}$ has the following structure:

$$\mathbf{A_l}(\mathbf{\theta_T}) = \left[\sqrt{\alpha_{Rl}^{T1}}\mathbf{a_l}(\theta_{T1}), \cdots, \sqrt{\alpha_{Rl}^{Tm}}\mathbf{a_l}(\theta_{TM}) \right] \tag{5}$$

where α_{Rl}^{Tm} is the propagation loss coefficient from the mth target to the lth array and is related to d_{Rl}^{Tm}. $\mathbf{a_l}(\theta_{Tm}) \in C^{P_l \times 1}$ is the direction vector from the mth target to the lth array. Under the assumption of a uniform linear array (ULA),

$$\mathbf{a_l}(\theta_{Tm}) = \left[1, e^{j\frac{2\pi\Delta_l}{\lambda}\sin\theta_{Tm}}, \cdots, e^{j(P_l-1)\frac{2\pi\Delta_l}{\lambda}\sin\theta_{Tm}} \right]^t \tag{6}$$

Under the first assumption, the matrix $\mathbf{B_l} \in C^{1 \times K}$ can be expressed as

$$\mathbf{B_l} = \left[\sqrt{\alpha_{Rl}^{J1}}e^{j\frac{2\pi d_{Rl}^{J1}}{\lambda}\sin\theta_{J1}}, \cdots, \sqrt{\alpha_{Rl}^{JK}}e^{j\frac{2\pi d_{Rl}^{JK}}{\lambda}\sin\theta_{JK}} \right] \tag{7}$$

where α_{Rl}^{Jk} is the propagation loss from the kth jammer to the lth array, which is related to d_{Rl}^{Jk}. In the case of a fixed receiver position (x_{Rl}, y_{Rl}), α_{Rl}^{Jk} depends on the position of the jammer (x_{Jk}, y_{Jk}).

Since the jamming noise and environmental noise received by the lth array are both Gaussian white noise and independent from each other, the superimposed noise is still Gaussian white noise, and the total power is the sum of the power of each noise source, which can be expressed as

$$\hat{\sigma}_l = \sum_{k=1}^{K} \alpha_{Rl}^{Jk} \sigma_k + \sigma_0 \tag{8}$$

where σ_k is the noise power emitted by the kth jammer and $\alpha_{Rl}^{Jk} \sigma_k$ is the noise power of the kth jammer received by the lth array.

Thus far, a receiving signal model of an UASN has been provided, which contains multiple targets and barrage jammers. The working principle of the barrage jamming method is to transmit high-power noise and reduce the signal-to-noise ratio of the UASN, thereby increasing its parameter estimation error. Therefore, the jamming effect depends on the noise power. According to Equations (2) and (8), the total power of the array receiving noise is closely related to the locations of the distributed jammers, especially in a sensor network containing multiple arrays. Therefore, the purpose of this paper is to study the optimal layout strategy for suppressing jammers of UASNs. In the next section, the CRB for the joint estimation of UASNs is presented to evaluate the jamming effect of distributed jammers.

3. Calculation of the CRB

The CRB is the best accuracy that a receiving system can achieve when estimating the target parameters. The CRB is usually used to measure the receiver performance and is used as a measure of the distributed jamming performance in this paper. In the case of constant target parameters, the higher the CRB of the receiver, the better the jamming effect. In this section, the CRB of the receiving sensor network for target angle estimation is derived, which considers the position of the jammers (x_{Jk}, y_{Jk}) as a variable. Based on this approach, the optimal layout strategy for the distributed jammers based on CRB maximization is presented in the next section. The CRB of a single sensor array with a Gaussian white noise background was given by Stoica and Nehorai [15]; however, this approach cannot be adopted directly in the model of this paper, which contains a 10-array network, where the receiving noise of each array is different but coherent [19] (the noise power of each jammer is superposed).

The CRB of the receiving system can be written as

$$\mathbf{CRB} = \left\{ \mathrm{E} \left[\frac{\partial^2 \ln L(y|\rho)}{\partial \rho^2} \right] \right\}^{-1} \tag{9}$$

where ρ is a collection of unknown parameters and $L(y|\rho)$ is the probability density function of ρ. In the model described above, the unknown estimated parameters of the receiving system are

$$\rho = \{\hat{\sigma}, \mathcal{R}(\hat{s}(t)), \mathcal{I}(\hat{s}(t)), \theta_T\} \tag{10}$$

where $\hat{\sigma}$ is the array receiving noise and is related to the position of the jammers (x_{Jk}, y_{Jk}), the emission noise power of the jammers σ_k, and the environmental noise power σ_0; θ_T is the arrival angle of the target signal, $\mathcal{R}(\hat{s}(t)) \triangleq \mathrm{Re}\hat{s}(t)$ is the real part of the target signal, and $\mathcal{I}(\hat{s}(t)) \triangleq \mathrm{Im}\hat{s}(t)$ is the imaginary part of the target signal.

The probability density function of joint estimation of multiple arrays can be written as

$$L(y|\rho) = \frac{1}{\pi^{MN} \det(\mathbf{Q})^{MN}} \exp\left\{ -\sum_{t=1}^{N} \left[\widetilde{\mathbf{y}}(t) - \widetilde{\mathbf{A}}\hat{s}(t) \right]^{H} \mathbf{Q}^{-1} \left[\widetilde{\mathbf{y}}(t) - \widetilde{\mathbf{A}}\hat{s}(t) \right] \right\} \tag{11}$$

where

$$\mathbf{Q} = \begin{bmatrix} \hat{\sigma}_1 & & \\ & \ddots & \\ & & \hat{\sigma}_L \end{bmatrix} \tag{12}$$

$$\widetilde{\mathbf{y}}(t) = [\mathbf{y}_1(t), \cdots, \mathbf{y}_\mathbf{L}(t)]^t \tag{13}$$

$$\widetilde{\mathbf{A}} = [\mathbf{A}_1(\boldsymbol{\theta}_\mathbf{T}), \cdots, \mathbf{A}_\mathbf{L}(\boldsymbol{\theta}_\mathbf{T})]^t \tag{14}$$

$$\mathbf{D}_\mathbf{l} = \frac{\partial \mathbf{A}_\mathbf{l}(\boldsymbol{\theta}_\mathbf{T})}{\partial \boldsymbol{\theta}_\mathbf{T}} \tag{15}$$

Then according to the derivation method used by Stoica and Nehorai [15], the CRB of the sensor network's estimation of the target angle can be obtained (see the detailed derivation process in Appendix A):

$$\mathbf{CRB}(\boldsymbol{\theta}_\mathbf{T}) = \left\{ \boldsymbol{\Gamma} - \sum_{t=1}^{N} \mathrm{Re}\left[\mathbf{V}_t^H \mathbf{G} \mathbf{V}_t\right] \right\}^{-1} \tag{16}$$

where

$$\boldsymbol{\Gamma} = \sum_{i=1}^{L}\sum_{j=1}^{L} \frac{2}{\hat{\sigma}_i \hat{\sigma}_j} \left(\sum_{k=1}^{K} \alpha_{Ri}^{Jk} \alpha_{Rj}^{Jk} \sigma_k \right) \sum_{t=1}^{N} \mathrm{Re}\left[\hat{s}^H(t) \mathbf{D_i}^H \mathbf{D_j} \hat{s}(t) \right] \tag{17}$$

$$\mathbf{V_t} = \sum_{i=1}^{L}\sum_{j=1}^{L} \frac{2}{\hat{\sigma}_i \hat{\sigma}_j} \left(\sum_{k=1}^{K} \alpha_{Ri}^{Jk} \alpha_{Rj}^{Jk} \sigma_k \right) \mathrm{Re}\left[\mathbf{A_i}^H \mathbf{D_j} \hat{s}(t) \right] \tag{18}$$

$$\mathbf{G} = \mathbf{H}^{-1} \tag{19}$$

$$\mathbf{H} = \sum_{i=1}^{L}\sum_{j=1}^{L} \frac{2}{\hat{\sigma}_i \hat{\sigma}_j} \left(\sum_{k=1}^{K} \alpha_{Ri}^{Jk} \alpha_{Rj}^{Jk} \sigma_k \right) \mathbf{A_i}^H \mathbf{A_j} \tag{20}$$

The CRB of the target angle estimation in Equation (15) is a function of all parameters of the target, receivers, jammers, and the environment. In this paper, the optimal layout strategy of the jammers is considered, thus the parameters of the target, receiver, and environment can be assumed to be fixed. Under this assumption, the CRB in Equation (15) varies only with the noise power emitted by each jammer σ_k and their locations (x_{Jk}, y_{Jk}). In practical situations, the available jammer resources are usually known; that is, the jammer transmit power is known. Under this condition, the CRB obtained in this paper is only related to the locations of the jammers. Therefore, by optimizing (x_{Jk}, y_{Jk}), the maximum CRB can be obtained, thereby obtaining the strongest jamming effect, which can reduce the estimation accuracy of the UASN.

4. Distributed Jamming Strategy Design

4.1. Optimization Model

According to the CRB obtained in the previous section, a distributed barrage jamming layout strategy for an UASN is proposed, which uses the CRB of the UASN as the evaluation index of jamming effect to find the optimal distribution of jammers, as shown in Figure 2. In this section, the distributed barrage jamming layout strategy is transformed into a mathematical model based on CRB maximization, where the locations of the jammers are used as a variable and the jamming performance is reflected by the jamming noise power.

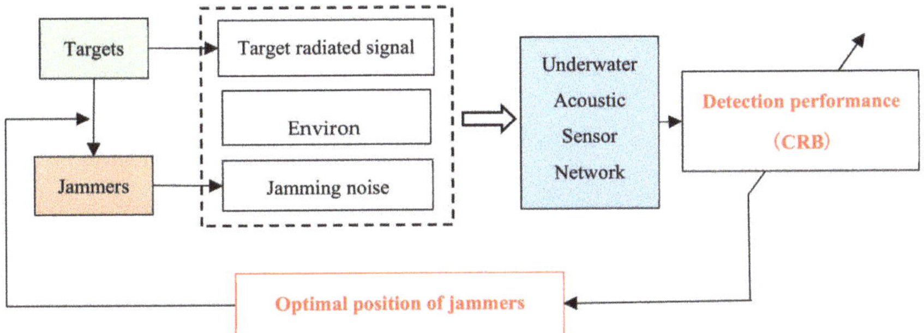

Figure 2. Model of optimal jamming layout strategy. CRB: CraméRao bound.

In the jamming scenario studied in this paper, a self-defense jamming strategy is considered, where the target signals are received by the receiver and the barrage jamming noise is transmitted to the receiver. Thus, the signal-to-noise ratio of the receiver is reduced, and the estimation error of the target is increased, which helps the target escape from the tracking of the UASN. Therefore, the signal of the receiver should include the target radiated signal, the emission noise of each jammer, and the environmental noise, as shown in Equation (4). A low-cost jammer is considered, which can radiate uniform jamming noise in all directions in space, so there is no need to use an array structure to modulate its phase. The use of such jammers can reduce the cost of the entire distributed jamming system, so more jammers can be employed to achieve better effects with complex sensor networks.

Since the jammers are launched by the target, the available jamming resources should be known when the jamming strategy is designed; that is, the number of jammers K, the emission jamming noise power σ_k, and the location of the jammers $\left(x_{Jk}, y_{Jk}\right)$ are known. Suppose that the jammer has obtained some basic information about the receiver through the detection equipment of the target, (e.g., the numbers of arrays L and the locations of the arrays $\left(x_{Rl}, y_{Rl}\right)$), then the CRB of the UASN can be obtained.

Under the assumptions in Section 2, the calculation of the total power of the received noise of each array is shown in Equation (8). The total received noise power of each array is the sum of the jamming noise power after propagation attenuation $\alpha_{Rl}^{Jk}\sigma_k$ and the environmental noise power σ_0. Since distributed jammers are usually distributed near the receivers, the propagation attenuation model of the jamming noise is considered as a spherical attenuation model in this paper. That is

$$\alpha_{Rl}^{Jk} = \frac{1}{\left(d_{Rl}^{Jk}\right)^2} = \frac{1}{\left(x_{Rl} - x_{Jk}\right)^2 + \left(y_{Rl} - y_{Jk}\right)^2} \tag{21}$$

Nevertheless, the propagation attenuation model of jamming noise depends on the actual marine environment and the jammer and receivers. Therefore, other propagation attenuation models can be used according to the actual situation when designing a distributed jamming layout strategy.

In Equation (21), α_{Rl}^{Jk} depends on the distance between the jammers and the receiver d_{Rl}^{Jk}. When the receiver distance is known, the total power of the receiver noise is only related to the locations of the jammers $\left(x_{Jk}, y_{Jk}\right)$. In polar coordinates, the locations can be expressed as $\left(r_k \sin\theta_{Jk}, r_k \cos\theta_{Jk}\right)$, where r_k is the distance from the kth jammer to the origin of the coordinates. Suppose the kth jammer is sent by the mth target; the geometry is shown in Figure 3. Usually, the distance between jammers and the target d_{Tm}^{Jk} is known; it is the distance travelled by the jammer from the target and always equal to the longest distance that the jammer can travel (to be closer to the receiver to get the maximum jamming performance). So d_{Tm}^{Jk} and r_k can be different for different jammers, and the latter can be calculated from the geometry in Equation (22):

$$\sqrt{\left(r_k \sin \theta_{Jk} - x_{Tm}\right)^2 + \left(r_k \cos \theta_{Jk} - y_{Tm}\right)^2} = d_{Tm}^{Jk} \tag{22}$$

Since r_k can be calculated from the known parameters, the angle of the jammer is the only variable in CRB (Equation (16)). Therefore, the optimal location layout strategy for the distributed jammers can be expressed as the optimal angle layout strategy, which depends on θ_{Jk}.

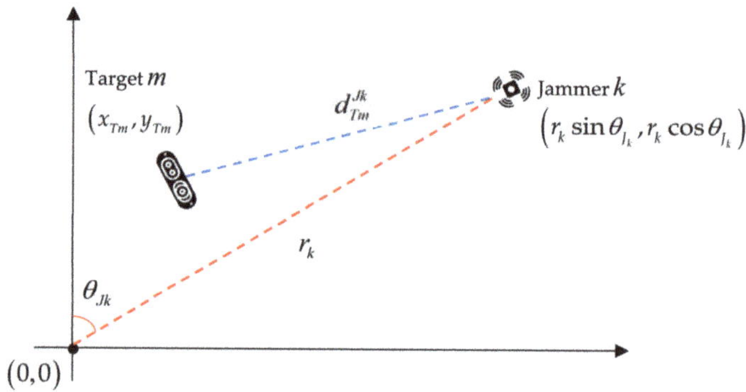

Figure 3. Geometry of the *m*th target and the *k*th jammer.

After clarifying the parameters of each component in the jamming scenario, an optimization model of the distributed jammer layout strategy based on CRB maximization is designed. The CRB of the UASN for the target angle estimation is given in Equation (16). For M targets, the resulting CRB is an $M \times M$ matrix, and the diagonal elements of the matrix are the estimated variances of the angles of each target.

$$\begin{cases} C_{\theta_{T1}} = \left[\mathbf{CRB}_{\theta_T}\right]_{1,1} \\ C_{\theta_{T2}} = \left[\mathbf{CRB}_{\theta_T}\right]_{2,2} \\ \quad \cdots \\ C_{\theta_{Tm}} = \left[\mathbf{CRB}_{\theta_T}\right]_{M,M} \end{cases} \tag{23}$$

By maximizing the weighted average CRB for the estimation of each target angle, the optimal angle of the jammers can be obtained

$$\begin{aligned} \max_{\theta_{J1}, \cdots, \theta_{J2}} \quad & \sum_{m=1}^{M} \lambda_m C_m \\ s.t. \quad & \hat{\sigma}_l = \sum_{k=1}^{K} \alpha_{Rl}^{Jk} \sigma_k + \sigma_0 \\ & \alpha_{Rl}^{Jk} = \frac{1}{\left(x_{Rl}-x_{Jk}\right)^2 + \left(y_{Rl}-y_{Jk}\right)^2} \\ & \left(x_{Jk}, y_{Jk}\right) = \left(r_k \sin \theta_{Jk}, r_k \cos \theta_{Jk}\right) \end{aligned} \tag{24}$$

where $\lambda_1, \cdots, \lambda_M$ are regularization factors that can be determined according to the importance of the corresponding target.

4.2. Problem Solution

The optimal layout strategy model for distributed jammers provided in this paper is based on a calculation of the CRB. Since the calculation formula of the CRB is very complicated, integrating various parameters of the receiving system and the received system, and highly nonconvex, heuristic algorithms can be used. Common heuristic algorithms include simulated annealing, greedy algorithms, tabu search, ant colony optimization, and genetic algorithms [20–25]. This paper uses particle swarm optimization (PSO) [24,25] to solve the given optimization problem. PSO is an iterative global optimization algorithm with simple parameters and an easy implementation, which is widely used in various complex nonconvex function optimization problems.

PSO is an evolutionary computing technology that was proposed by Eberhart and Kennedy in 1995 [25]. PSO simulates the flight and foraging behavior of bird swarms; it is a simplified model based on swarm intelligence. Each optimization problem solution is imagined as a bird, called a particle. All particles are searched in a given D-dimensional space, and the search fitness value is calculated by a given fitness function to judge the current search position $\mathbf{x}_i^{(n)}$, where $\mathbf{x}_i^{(n)} = \left[x_{i1}^{(n)}, \cdots, x_{iD}^{(n)} \right]$. Each particle records the best position $\mathbf{pbest}_i^{(n)}$ searched so far and uses a speed $\mathbf{v}_i^{(n)}$ to determine the direction and step size of the next iteration, where $\mathbf{pbest}_i^{(n)} = \left[pbest_{i1}^{(n)}, \cdots, pbest_{iD}^{(n)} \right]$ and $\mathbf{v}_i^{(n)} = \left[v_{i1}^{(n)}, \cdots, v_{iD}^{(n)} \right]$. The fitness values of all particles are compared to obtain the best position acquired by the population $\mathbf{gbest}^{(n)}$, where $\mathbf{gbest}^{(n)} = \left[gbest_1^{(n)}, \cdots, gbest_d^{(n)} \right]$. The iteration formulas of the speed in each dimension $v_{id}^{(n)}$ and the position in each dimension $x_{id}^{(n)}$ in the PSO algorithm are

$$v_{id}^{(n)} = w v_{id}^{(n-1)} + c_1 r_1 \left(pbest_{id} - x_{id}^{(n-1)} \right) + c_2 r_2 \left(gbest_d^{(n-1)} - x_{id}^{(n-1)} \right)$$
$$x_{id}^{(n)} = x_{id}^{(n-1)} + v_{id}^{(n-1)}$$

(25)

where c_1 and c_2 are the acceleration constants, r_1 and r_2 are random functions, and w is the inertia weight. The specific parameter settings are described in detail by Kennedy and Eberhart [25] and Poli and colleagues [26].

In the optimization model of this paper, the final solution should be the angles of K jammers. Therefore, the dimension of each particle should be set as $D = K$, and the position of each particle should be set as $x_{id} = \theta_{J_k}$. According to the block diagram of the PSO algorithm (Figure 4), the optimal location of each jammer can be obtained.

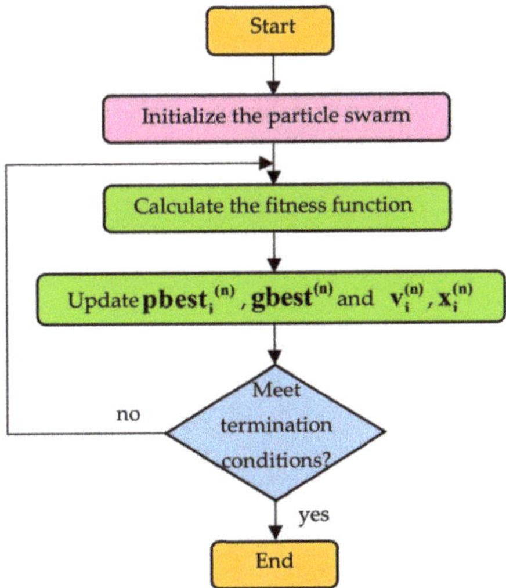

Figure 4. Flow chart of particle swarm optimization (PSO).

5. Numerical Experiments

In this section, the effectiveness of the proposed jamming strategy is analyzed through numerical simulation experiments. The jamming scenario in this paper includes an UASN (consisting of three ULAs), targets, distributed jammers, and environmental noise. Therefore, different array parameters, environmental noise, and target parameters are considered to study their impact on the optimal placement angle of the distributed jammers. In addition, different layout strategies are compared, including (1) concentrated (all of the jammers are concentrated using the array angle with better detection performance); (2) random (the jammers are randomly distributed in all array directions); (3) uniform (the jammers are distributed between array angles of arrays at equal intervals); and (4) optimized (the proposed layout strategy in this paper) layouts.

5.1. Influence of the Sensor Networks

First, the influence of the sensor network on the optimal distribution angle should be considered. The array angle, the distance between the array and the target, and the number of array elements in each array are set as variables, because all three parameters will affect the target detection performance of the sensor network. In the simulations in this section, a single target scenario is considered in which the target's location is set to $(0, 0)$, and the target radiated signal is assumed to be Gaussian white noise with a power of 110 dB. Meanwhile, three arrays, whose normal directions point to the target, and the environmental noise received by the array elements, which is Gaussian white noise with a power of 50 dB, compose the sensor network at the same time. The jammer, which is 2000 m away from the target, emits Gaussian white noise with a power of 110 dB. The tables below show the results of the following three simulations under different scenarios:

(1) The distribution angle (relative to the target) of each array is used as a variable. The distance of each array to the target is 2500 m, and all the arrays are eight-element ULAs;
(2) The distance of each array to the target is used as a variable. The angles of the arrays are $10°$, $30°$, and $50°$, and all the arrays are eight-element ULAs;
(3) The element numbers of each array are used as a variable. The distance of each array to the target is 2500 m, the angles of the arrays are $10°$, $30°$, and $50°$, and all the arrays are eight-element ULAs.

The first simulation of this section assumes that each array in the sensor network has the same array structure and the same distance to the target, which means that every array has the same target detection performance. From the calculation results of the three jammers in Table 2, it can be seen that when the detection performance of each array is the same, the jammers tend to be placed between each array and the target. The calculation results of the six jammers indicate the same conclusion. When the jammers cannot be evenly distributed among the array element angles (the number of jammers cannot be divided by the number of arrays), such as the case of four jammers, it can be found that three of the jammers are still distributed on the lines between each array and the target. The other jammer can be placed on any of the array elements randomly.

Table 2. Optimized angles of the jammers for different numbers of jammers with respect to the three-array networks with different angles in a single-target scene.

Angles of the Arrays (in °)	Angles of the Jammers (in °)		
3 arrays	3 jammers	4 jammers	6 jammers
(10, 30, 50)	(11.8, 29.9, 48.4)	(11.0, 29.8, 30.1, 48.7)	(11.8, 12.3, 29.8, 29.9, 47.8, 48.4)
(10, 40, 70)	(10.6, 40.0, 69.0)	(10.5, 40.0, 40.1, 69.4)	(10.9, 11.3, 39.8, 39.9, 68.9, 69.2)
(0, 20, 60)	(2.1, 18.2, 59.6)	(1.9, 17.5, 58.6, 59.3)	(2.0, 2.3, 17.9, 18.2, 59.5, 59.7)
(0, 20, 80)	(2.5, 17.9, 80.0)	(1.8, 17.7, 79.5, 79.7)	(2.0, 2.1, 17.9, 18.8, 79.9, 80.4)

The second simulation of this section assumes that each array in the UASN has the same array structure and fixed angles, so the closer the array is to the target, the better the detection performance.

From the calculation results of the three jammers in Table 3, it can be seen that when the differences among the distances of the arrays are small, the jammers still tend to be distributed in the same way as the array angles. However, there also exists a tendency for the jammers to gather at arrays that have stronger detection performance. The tendency is more pronounced when the differences among the distances of the arrays are larger. When the difference reaches a certain level, all of the jammers will focus on the array with the strongest detection performance, which means that the jamming system is equivalent to a centralized jamming system. The case with four jammers produces similar results, and when the detection performance of each array is different, the remaining jammers after the equalization will focus on the array angle with the strongest detection performance.

Table 3. Optimized angles of the jammers for different numbers of jammers with respect to the three-array networks with different distances in a single-target scene.

Angles of the Arrays (in °)	Distances of Arrays (in m)	Angles of the Jammers (in °)	
3 arrays		3 jammers	4 jammers
(10, 30, 50)	(2600, 2500, 2600)	(13.1, 30.0, 46.8)	(11.9, 29.9, 30.4, 47.8)
	(2700, 2500, 2700)	(15.8, 29.7, 44.5)	(13.4, 30.1, 30.1, 46.7)
	(2700, 2500, 2900)	(15.1, 31.7, 31.9)	(13.6, 31.2, 31.5, 31.7)
	(2900, 2500, 2900)	(29.8, 30.0, 30.1)	(29.6, 30.0, 30.0, 30.2)

The third simulation of this section assumes that the distance from each array to the target in the UASN is equal and the angles are fixed. The more elements the array contains, the better the detection performance. The results in Table 4 are similar to those in Table 3. As the difference in the array detection performance increases, the angles of the jammers tend to be distributed evenly across the arrays and tend to be concentrated at arrays with the highest performance.

Table 4. Optimized angles of the jammers for different numbers of jammers with respect to the three-array networks with different numbers of elements in a single-target scene.

Angles of the Arrays (in °)	Numbers of Elements	Angles of the Jammers (in °)	
3 arrays		3 jammers	4 jammers
(10, 30, 50)	(8, 10, 8)	(14.2, 30.2, 46.1)	(12.3, 29.7, 30.0, 48.1)
	(8, 11, 8)	(16.4, 30.1, 43.4)	(13.3, 29.9, 30.3, 46.9)
	(8, 11, 10)	(27.3, 27.5, 48.1)	(13.3, 31.1, 31.4, 48.5)
	(8, 12, 8)	(30.0, 30.0, 30.1)	(27.0, 27.3, 27.4, 46.0)

5.2. Influence of Environmental Noise

This section considers the influence of environmental noise on the optimal angles of distributed jammers, and the simulation scenario is the same as in Section 5.1. The UASN consists of three 8-element ULAs, and the angles of the array relative to the target are fixed to $(10°, 30°, 50°)$. Assuming that the distance from each array to the target is fixed at (2600 m, 2500 m, 2600 m), in this sensor network, the array at $30°$ has the strongest detection performance. In the second case, the distance from each array to the target is fixed to (2500 m, 2500 m, 2500 m), and then the receiving performance of each array in this sensor network is the same. Table 5 shows the results of the optimal jamming angles under environmental noise with different values of the variance σ_0. Another case adopts the centralized layout strategy, random layout strategy, uniform layout strategy, and optimal layout strategy under environmental noise with different variances. The results of the second case are all indicated in Figure 5.

Table 5. Optimized angles of three jammers with respect to the three-array networks under environmental noise with different variances in a single-target scene.

Environmental Noise Variance σ_0 (in dB)	Angles of the Jammers in the First Situation (in °)	Angles of the Jammers in the Second Situation (in °)
40	(12.9, 29.8, 47.3)	(11.1, 30.0, 48.1)
50	(13.2, 29.9, 46.9)	(11.8, 30.2, 48.2)
60	(29.9, 30.1, 30.2)	(12.4, 30.3, 47.8)
70	(29.9, 30.0, 30.2)	(30.1, 30.1, 30.2)

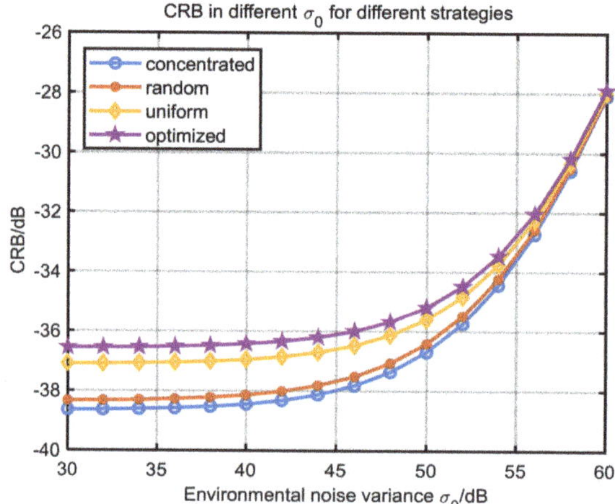

Figure 5. Cramér–Rao bound (CRB) under environmental noise with different variances for different strategies in a single-target scene.

In the first case, the simulation assumes that the array with the 30° direction is closest to the target, so it has the strongest detection performance. Table 5 shows that when the variance of the environmental noise is small, the optimal angles of the jammers are consistent with the conclusion in Section 5.1 (i.e., they are close to the array angles, and when the variance of the environmental noise gradually increases, the optimal angles of the jammers tend to become compact). The second case shows that regardless of whether the array detection performance is the same, the optimal angle of each jammer is concentrated in the array in the middle of the sensor network as long as the environmental variance is large enough.

Figure 5 indicates that the proposed optimization strategy has a higher CRB of the target angle estimation than the other three strategies, especially in the case of low environmental noise. The CRBs obtained by the four strategies are not very different under high environmental noise, because the noise signal affecting the receiver in this case is mainly environmental noise. Therefore, the proposed optimal distributed jamming layout strategy has greater advantages in the case of low environmental noise.

5.3. Multitarget Scene

In this section, the optimal layout strategy for distributed jammers in a multitarget scenario is considered, which contains three targets with the same power as the radiation signal and are located at (0,0), (2000,0), and (0,1000) (the unit is m). Assuming that every target is equally important, the regularization factor in Equation (23) is $\lambda_1 = \lambda_2 = \lambda_3 = 1/3$. Similarly, three eight-element ULAs are used. The angle of each array with respect to the target is fixed at (10°, 30°, 50°), and the distance from each array to the target is fixed at (2500 m, 2500 m, 2500 m). Six jammers with an emission noise power of 110 dB are used. Based on the jamming scenario in Figure 1, the optimal angles for $\sigma_0 = 30$ dB are indicated in Figure 6. Then, in the case of $\sigma_0 = 30 \sim 60$ dB, four different jamming strategies are adopted, and the CRBs are indicated in Figure 7.

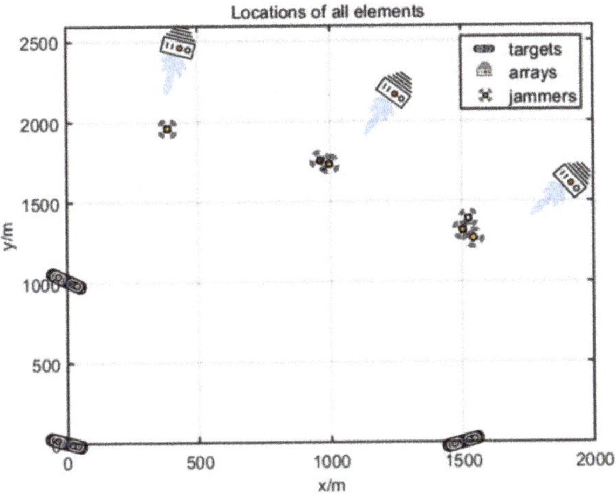

Figure 6. Locations of the elements in the model of a multitarget scene.

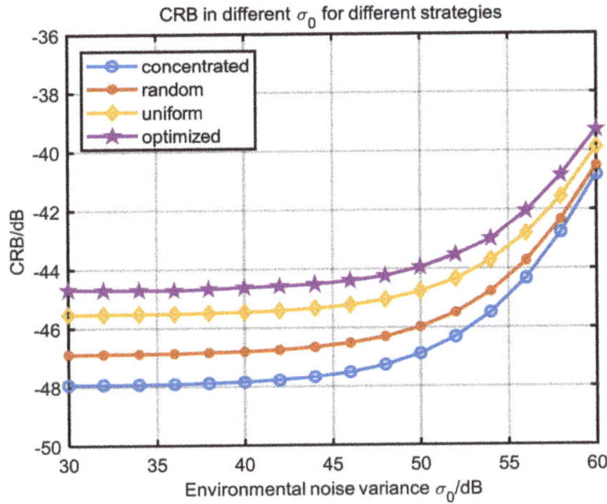

Figure 7. CRB under different environmental noise for different strategies in a multitarget scene.

Figure 6 shows that in the case of multiple targets, the jammers still tend to be distributed according to the array angles. Since distributed jamming is usually a short-range jamming method, the optimal distribution of the angles of the jammers has a greater relationship with the structure of the sensor network. Figure 7 shows the CRB of the target angle estimation of the sensor network obtained by using the concentrated layout strategy, random layout strategy, uniform layout strategy, and optimal layout strategy under different environmental noise conditions; from the results, we find that the proposed strategy has the highest CRB among these strategies, which means it has the strongest impact on the receiver's parameter estimation performance. That is, the proposed optimal strategy can also achieve better jamming performance than the other strategies in a multitarget scenario.

5.4. Analysis of the Position Error

There is strong turbulence in a real marine environment, which can have a non-negligible effect on the position of an underwater combat platform. Thus, the real position of the jammers will deviate greatly from the original position. In this section, the location error of the jammers is discussed to verify the effectiveness of the proposed optimized layout strategy. The multitarget scenario is considered, and the parameters of the targets and UASN in this section are the same as those in Section 5.3. Assuming that the distance error of the jammer caused by factors such as ocean turbulence and wind follows a normal distribution with a mean of 0 and a standard deviation of 30 m, the angle error of the jammer follows a normal distribution with a mean of 0 and a standard deviation of 3°. Due to the errors, the actual position of the jammer will fluctuate around the calculated value, so the CRB calculated from the actual position will also fluctuate near the theoretical value. Figure 8 shows the possible locations of jammers for $\sigma_0 = 30$ dB in the presence of errors. The fluctuation range of the CRB obtained by the proposed optimized layout strategy in the presence of location errors is indicated in Figure 9. Since the uniform layout strategy achieves better jamming performance than the random layout strategy and concentrated layout strategy, the uniform layout strategy is used for a comparison with the optimized strategy.

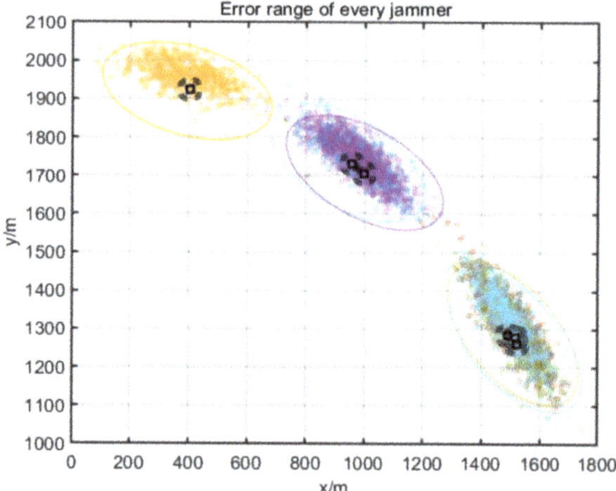

Figure 8. Range of the real locations of the jammers.

Figure 8 shows obvious error in the location of the jammers compared with the calculated optimal location, and Figure 9 presents the calculated CRB under this error condition. It is obvious that the proposed optimal layout strategy achieves a better performance than the uniform layout strategy. The lower bound of the CRB fluctuation range caused by the position error of the jammers with the optimized strategy is higher than the upper bound of the CRB fluctuation range with the uniform strategy for most environmental noise conditions. The CRB fluctuation range is roughly the same for both the optimized strategy and the uniform strategy, so it can be inferred that for the random layout strategy and the centralized layout strategy, the CRB fluctuation range is also roughly the same and lower. Thus, the proposed optimal strategy still achieves the best jamming performance when there is a deviation in the real location of the jammers. It can also be found that as the environmental noise increases, the fluctuation in the CRB caused by the position error of the jammers decreases, which means that the real position of the jammers will not affect the jamming effect. This is because the main barrage effect for the UASN comes from environmental noise in this case. From the experiment in this section, a situation close to the real marine environment is considered. The result has proven that the optimal strategy proposed in this paper still achieves a significantly better jamming performance even in the case of uncertain errors, which verifies its practicability.

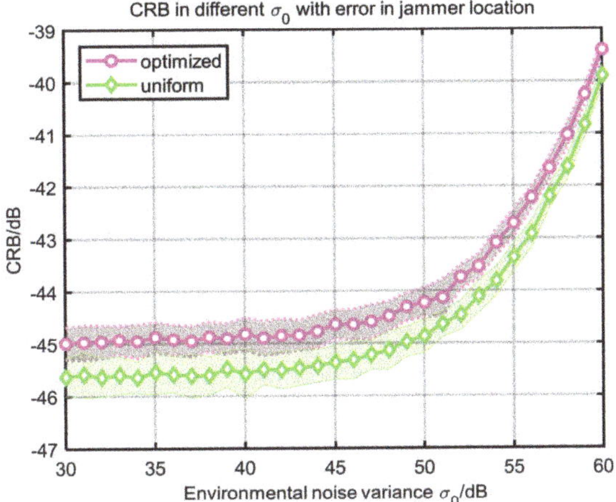

Figure 9. CRB under environmental noise with different variances.

6. Conclusions

In this paper, the optimal layout strategy for distributed barrage jammers against UASNs is studied. The CRB of UASNs estimating target angles in the case of multiple targets and multiple jammers is derived, which uses the angle of the jammers as a variable. Based on this result, a distributed jamming optimization strategy is proposed, which uses the maximum CRB as the cost function and aims to obtain the optimal distribution of the jammers. PSO algorithm is used to solve this complex and nonconvex model. Numerical experiments are executed to compare the jamming strategy with three traditional strategies. The results show that the proposed optimal layout strategy for distributed jammers can achieve stronger jamming effects than the other strategies. The position error of the jammers caused by turbulence and a hurricane in a real ocean is considered, and the result shows that the proposed optimal layout strategy still performs better than the other strategies under obvious error conditions.

Future research will focus on more efficient jamming strategies to adapt to complex real-world scenarios, including using the mix jamming method of barrage jammers and deceptive jammers and increasing the speed of algorithm operations.

Author Contributions: Conceptualization, J.Z. and Y.D.; methodology, M.X. and J.Z.; software, M.X. and X.J.; investigation, Y.D.; writing—original draft preparation, M.X. and X.J.; writing—review and editing, M.X., J.Z., Y.D. and X.J.; supervision, J.Z. and Y.D.; project administration, J.Z.; funding acquisition, M.X. and J.Z. All authors have read and agreed to the published version of the manuscript.

Funding: This research was sponsored by the National Key R&D Program of China, No. 2016YFC1400104, and the Seed Foundation of Innovation and Creation for Graduate Students in Northwestern Polytechnical University, No. ZZ2019005.

Acknowledgments: We appreciate all editors and reviewers for the valuable suggestions and constructive comments. Thank you to everyone who contributed to this article.

Conflicts of Interest: The authors declare no conflicts of interest.

Appendix A

As shown in Equation (11), the joint likelihood function of the data received by UASNs is

$$L(y|\rho) = \frac{1}{\pi^{MN}\det(\mathbf{Q})^{MN}} \exp\left\{-\sum_{t=1}^{N}\left[\widetilde{\mathbf{y}}(t) - \widetilde{\mathbf{A}}\hat{\mathbf{s}}(t)\right]^{H}\mathbf{Q}^{-1}\left[\widetilde{\mathbf{y}}(t) - \widetilde{\mathbf{A}}\hat{\mathbf{s}}(t)\right]\right\} \tag{A1}$$

Thus, the log-likelihood function is

$$\ln L = \text{const} - mN\ln\prod_{l=1}^{L}\hat{\sigma}_l - \sum_{t=1}^{N}\sum_{l=1}^{L}\frac{1}{\hat{\sigma}_l}[\mathbf{y}_l(t) - \mathbf{A}_l\hat{\mathbf{s}}(t)]^{H}[\mathbf{y}_l(t) - \mathbf{A}_l\hat{\mathbf{s}}(t)] \tag{A2}$$

Denote $\mathbf{g}_l(t) = \mathbf{B}_l\hat{\mathbf{n}}(t) + e(t)$, as the total noise received by the lth sensor array. The derivatives of Equation (A2) with respect to $\mathcal{R}(\hat{\mathbf{s}}(t))$, $\mathcal{I}(\hat{\mathbf{s}}(t))$, and θ_T are

$$\frac{\partial \ln L}{\partial \mathcal{R}(\hat{\mathbf{s}}(t))} = \sum_{l=1}^{L}\frac{2}{\hat{\sigma}_l}\text{Re}\left[\mathbf{A}_l^{H}\mathbf{g}_l(t)\right] \quad t = 1,\cdots,N \tag{A3}$$

$$\frac{\partial \ln L}{\partial \mathcal{I}(\hat{\mathbf{s}}(t))} = \sum_{l=1}^{L}\frac{2}{\hat{\sigma}_l}\text{Im}\left[\mathbf{A}_l^{H}\mathbf{g}_l(t)\right] \quad t = 1,\cdots,N \tag{A4}$$

$$\frac{\partial \ln L}{\partial \theta_T} = \sum_{l=1}^{L}\frac{2}{\hat{\sigma}_l}\sum_{t=1}^{N}\text{Re}\left[\hat{\mathbf{s}}^{H}(t)\mathbf{D}_l^{H}\mathbf{g}_l(t)\right] \quad t = 1,\cdots,N \tag{A5}$$

with

$$\text{Re}(x)\text{Re}(y^t) = \frac{1}{2}\left\{\left[\text{Re}(xy^t) + \text{Re}(xy^{H})\right]\right\}$$

$$\text{Im}(x)\text{Im}(y^t) = -\frac{1}{2}\left[\text{Re}(xy^t) - \text{Re}(xy^{H})\right]$$

$$\text{Re}(x)\text{Im}(y^t) = \frac{1}{2}\left[\text{Im}(xy^t) - \text{Im}(xy^{H})\right]$$

we have

$$E\left[\frac{\partial \ln L}{\partial \mathcal{R}(\hat{\mathbf{s}}(m))}\right]\left[\frac{\partial \ln L}{\partial \mathcal{R}(\hat{\mathbf{s}}(n))}\right]^{t} = \sum_{i=1}^{L}\sum_{j=1}^{L}\frac{2}{\hat{\sigma}_i\hat{\sigma}_j}\left(\sum_{k=1}^{K}\alpha_{Ri}^{Jk}\alpha_{Rj}^{Jk}\sigma_k\right)\text{Re}\left[\mathbf{A}_i^{H}\mathbf{A}_j\right]\delta_{m,n} \tag{A6}$$

$$E\left[\frac{\partial \ln L}{\partial \mathcal{R}(\hat{\mathbf{s}}(m))}\right]\left[\frac{\partial \ln L}{\partial \mathcal{I}(\hat{\mathbf{s}}(n))}\right]^{t} = -\sum_{i=1}^{L}\sum_{j=1}^{L}\frac{2}{\hat{\sigma}_i\hat{\sigma}_j}\left(\sum_{k=1}^{K}\alpha_{Ri}^{Jk}\alpha_{Rj}^{Jk}\sigma_k\right)\text{Re}\left[\mathbf{A}_i^{H}\mathbf{A}_j\right]\delta_{m,n} \tag{A7}$$

$$E\left[\frac{\partial \ln L}{\partial \mathcal{R}(\hat{\mathbf{s}}(t))}\right]\left[\frac{\partial \ln L}{\partial \theta_T}\right]^{t} = \sum_{i=1}^{L}\sum_{j=1}^{L}\frac{2}{\hat{\sigma}_i\hat{\sigma}_j}\left(\sum_{k=1}^{K}\alpha_{Ri}^{Jk}\alpha_{Rj}^{Jk}\sigma_k\right)\text{Re}\left[\mathbf{A}_i^{H}\mathbf{D}_j\hat{\mathbf{s}}(t)\right] \tag{A8}$$

$$E\left[\frac{\partial \ln L}{\partial \mathcal{I}(\hat{\mathbf{s}}(m))}\right]\left[\frac{\partial \ln L}{\partial \mathcal{I}(\hat{\mathbf{s}}(n))}\right]^{t} = \sum_{i=1}^{L}\sum_{j=1}^{L}\frac{2}{\hat{\sigma}_i\hat{\sigma}_j}\left(\sum_{k=1}^{K}\alpha_{Ri}^{Jk}\alpha_{Rj}^{Jk}\sigma_k\right)\text{Re}\left[\mathbf{A}_i^{H}\mathbf{A}_j\right]\delta_{m,n} \tag{A9}$$

$$E\left[\frac{\partial \ln L}{\partial \theta_T}\right]\left[\frac{\partial \ln L}{\partial \theta_T}\right]^{t} = \sum_{i=1}^{L}\sum_{j=1}^{L}\frac{2}{\hat{\sigma}_i\hat{\sigma}_j}\left(\sum_{k=1}^{K}\alpha_{Ri}^{Jk}\alpha_{Rj}^{Jk}\sigma_k\right)\sum_{t=1}^{N}\text{Re}\left[\hat{\mathbf{s}}(t)^{H}\mathbf{D}_i^{H}\mathbf{D}_j\hat{\mathbf{s}}(t)\right] \tag{A10}$$

where $m = 1,\cdots,N$ and $n = 1,\cdots,N$. $\delta_{m,n}$ is the Dirac delta function ($\delta_{m,n} = 1$ if m = n and $\delta_{m,n} = 0$ otherwise).

Introduce the following notations

$$\boldsymbol{\Gamma} = \sum_{i=1}^{L}\sum_{j=1}^{L}\frac{2}{\hat{\sigma}_i\hat{\sigma}_j}\left(\sum_{k=1}^{K}\alpha_{Ri}^{Jk}\alpha_{Rj}^{Jk}\sigma_k\right)\sum_{t=1}^{N}\mathrm{Re}\left[\hat{s}^{\mathrm{H}}(t)\mathbf{D_i}^{\mathrm{H}}\mathbf{D_j}\hat{s}(t)\right] \tag{A11}$$

$$\mathbf{V_t} = \sum_{i=1}^{L}\sum_{j=1}^{L}\frac{2}{\hat{\sigma}_i\hat{\sigma}_j}\left(\sum_{k=1}^{K}\alpha_{Ri}^{Jk}\alpha_{Rj}^{Jk}\sigma_k\right)\mathrm{Re}\left[\mathbf{A_i}^{\mathrm{H}}\mathbf{D_j}\hat{s}(t)\right] \tag{A12}$$

$$\mathbf{G} = \mathbf{H}^{-1} \tag{A13}$$

$$\mathbf{H} = \sum_{i=1}^{L}\sum_{j=1}^{L}\frac{2}{\hat{\sigma}_i\hat{\sigma}_j}\left(\sum_{k=1}^{K}\alpha_{Ri}^{Jk}\alpha_{Rj}^{Jk}\sigma_k\right)\mathbf{A_i}^{\mathrm{H}}\mathbf{A_j} \tag{A14}$$

The analysis of the CRB matrix is the same as in the study by Xiao and colleagues [14], so the CRB can be expressed as

$$\mathbf{CRB}(\theta) = \left\{\boldsymbol{\Gamma} - \sum_{t=1}^{N}\mathrm{Re}\left[\mathbf{V}_t^{\mathrm{H}}\mathbf{G}\mathbf{V}_t\right]\right\}^{-1} \tag{A15}$$

References

1. Heidemann, J.; Stojanovic, M.; Zorzi, M. Underwater sensor networks: applications, advances and challenges. *Philos. Trans. R. Soc. A Math. Phys. Eng. Sci.* **2012**, *370*, 158–175. [CrossRef] [PubMed]
2. Faheem, M.; Butt, R.A.; Raza, B.; Alquhayz, H.; Ashraf, M.W.; Shah, S.B.; Ngadi, M.A.; Gungor, V.C. QoSRP: A Cross-Layer QoS Channel-Aware Routing Protocol for the Internet of Underwater Acoustic Sensor Networks. *Sensors* **2019**, *19*, 4762. [CrossRef] [PubMed]
3. Sendra, S.; Lloret, J.; Jimenez, J.M.; Parra, L. Underwater Acoustic Modems. *IEEE Sens. J.* **2016**, *16*, 4063–4071. [CrossRef]
4. Chandrasekhar, V.; Seah, W.K.; Choo, Y.S.; Ee, H.V. Localization in underwater sensor networks: Survey and challenges. In Proceedings of the 1st ACM international workshop on Underwater networks, Los Angeles, CA, USA, 21–25 September 2006; pp. 33–40.
5. Misra, S.; Dash, S.; Khatua, M.; Vasilakos, A.V.; Obaidat, M.S. Jamming in underwater sensor networks: detection and mitigation. *IET Commun.* **2012**, *6*, 2178–2188. [CrossRef]
6. Aphek, O.; Tidhar, G.; Goichman, T. Distributed Jammer System. U.S. Patent No. 7,920,255, 5 April 2011.
7. Song, X.; Willett, P.; Zhou, S.; Luh, P.B. The MIMO Radar and Jammer Games. *IEEE Trans. Signal Process.* **2012**, *60*, 687–699. [CrossRef]
8. Deligiannis, A.; Rossetti, G.; Panoui, A.; Lambotharan, S.; Chambers, J.A. Power allocation game between a radar network and multiple jammers. In Proceedings of the 2016 IEEE Radar Conference (RadarConf), IEEE, Philadelphia, PA, USA, 2–6 May 2016; pp. 1–5.
9. Zheng, G.; Na, S.; Huang, T.; Wang, L. A barrage jamming strategy based on CRB maximization against distributed MIMO radar. *Sensors* **2019**, *19*, 2453. [CrossRef] [PubMed]
10. Kashyap, A.; Basar, T.; Srikant, R. Correlated jamming on MIMO Gaussian fading channels. *IEEE Trans. Inf. Theory* **2004**, *50*, 2119–2123. [CrossRef]
11. Chavali, P.; Nehorai, A. Scheduling and power allocation in a cognitive radar network for multiple-target tracking. *IEEE Trans. Signal Process.* **2011**, *60*, 715–729. [CrossRef]
12. Akyildiz, I.F.; Pompili, D.; Melodia, T. Underwater acoustic sensor networks: research challenges. *Ad Hoc Netw.* **2005**, *3*, 257–279. [CrossRef]
13. Vadori, V.; Scalabrin, M.; Guglielmi, A.V.; Badia, L. Jamming in underwater sensor networks as a Bayesian zero-sum game with position uncertainty. In Proceedings of the 2015 IEEE Global Communications Conference (GLOBECOM), IEEE, San Diego, CA, USA, 6–10 December 2015; pp. 1–6.
14. Xiao, L.; Li, Q.; Chen, T.; Cheng, E.; Dai, H. Jamming games in underwater sensor networks with reinforcement learning. In Proceedings of the 2015 IEEE Global Communications Conference (GLOBECOM), IEEE, San Diego, CA, USA, 6–10 December 2015; pp. 1–6.

15. Stoica, P.; Nehorai, A. MUSIC, maximum likelihood, and Cramer-Rao bound. *IEEE Trans. Acoust. Speech Signal Process.* **1989**, *37*, 720–741. [CrossRef]

16. Yip, L.; Chen, J.C.; Hudson, R.E.; Yao, K. Cramer-Rao bound analysis of wideband source localization and DOA estimation. In Proceedings of the Advanced Signal Processing Algorithms, Architectures, and Implementations XII, International Society for Optics and Photonics, Seattle, WA, USA, 6 December 2002; Volume 4791, pp. 304–316.

17. Vu, P.; Haimovich, A.M.; Himed, B. Bayesian Cramer-Rao Bound for multiple targets tracking in MIMO radar. In Proceedings of the 2017 IEEE Radar Conference (RadarConf), IEEE, Seattle, WA, USA, 8–12 May 2017; pp. 938–942.

18. Chen, J.C.; Hudson, R.E.; Yao, K. Maximum-likelihood source localization and unknown sensor location estimation for wideband signals in the near-field. *IEEE Trans. Signal Process.* **2002**, *50*, 1843–1854. [CrossRef]

19. Lu, L.; Wu, H.C.; Yan, K.; Iyengar, S.S. Robust expectation-maximization algorithm for multiple wideband acoustic source localization in the presence of nonuniform noise variances. *IEEE Sens. J.* **2010**, *11*, 536–544. [CrossRef]

20. Siddique, N.; Adeli, H. Simulated annealing, its variants and engineering applications. *Int. J. Artif. Intell. Tools* **2016**, *25*, 1630001. [CrossRef]

21. Wang, Y.; Cong, G.; Song, G.; Xie, K. Community-based greedy algorithm for mining top-k influential nodes in mobile social networks. In Proceedings of the 16th ACM SIGKDD international conference on Knowledge discovery and data mining, Association for Computing Machinery, New York, NY, USA, 24–28 July 2010; pp. 1039–1048.

22. Smutnicki, C.; Bożejko, W. Tabu search and solution space analyses. The job shop case. In *International Conference on Computer Aided Systems Theory*; Springer: Cham, Switzerland, 2017; pp. 383–391.

23. Yang, Q.; Chen, W.N.; Yu, Z.; Gu, T.; Li, Y.; Zhang, H.; Zhang, J. Adaptive multimodal continuous ant colony optimization. *IEEE Trans. Evol. Comput.* **2016**, *21*, 191–205. [CrossRef]

24. Wang, D.; Tan, D.; Liu, L. Particle swarm optimization algorithm: an overview. *Soft Comput.* **2018**, *22*, 387–408. [CrossRef]

25. Kennedy, J.; Eberhart, R. Particle swarm optimization. In Proceedings of the ICNN'95-International Conference on Neural Networks, IEEE, Perth, WA, Australia, 27 November–1 December 1995; Volume 4, pp. 1942–1948.

26. Poli, R.; Kennedy, J.; Blackwell, T. Particle swarm optimization. *Swarm Intell.* **2007**, *1*, 33–57. [CrossRef]

Journal of
Marine Science and Engineering

Article

Robust Capon Beamforming against Steering Vector Error Dominated by Large Direction-of-Arrival Mismatch for Passive Sonar

Yu Hao [1,2,3], Nan Zou [1,2,3,4,*] and Guolong Liang [1,2,3,4]

1   Acoustic Science and Technology Laboratory, Harbin Engineering University, Harbin 150001, China; haoyu@hrbeu.edu.cn (Y.H.); liangguolong@hrbeu.edu.cn (G.L.)
2   Key Laboratory of Marine Information Acquisition and Security (Harbin Engineering University), Ministry of Industry and Information Technology, Harbin 150001, China
3   College of Underwater Acoustic Engineering, Harbin Engineering University, Harbin 150001, China
4   Qingdao Haina Underwater Info. Tech. Co. Ltd., Qingdao 266400, China
*   Correspondence: zounan@hrbeu.edu.cn; Tel.: +86-151-2450-3518

Received: 25 February 2019; Accepted: 22 March 2019; Published: 26 March 2019

Abstract: Capon beamforming is often applied in passive sonar to improve the detectability of weak underwater targets. However, we often have no accurate prior information of the direction-of-arrival (DOA) of the target in the practical applications of passive sonar. In this case, Capon beamformer will suffer from performance degradation due to the steering vector error dominated by large DOA mismatch. To solve this, a new robust Capon beamforming approach is proposed. The essence of the proposed method is to decompose the actual steering vector into two components by oblique projection onto a subspace and then estimate the actual steering vector in two steps. First, we estimate the oblique projection steering vector within the subspace by maximizing the output power while controlling the power from the sidelobe region. Subsequently, we search for the actual steering vector within the neighborhood of the estimated oblique projection steering vector by maximizing the output signal-to-interference-plus-noise ratio (SINR). Semidefinite relaxation and Charnes-Cooper transformation are utilized to derive convex formulations of the estimation problems, and the optimal solutions are obtained by the rank-one decomposition theorem. Numerical simulations demonstrate that the proposed method can provide superior performance, as compared with several previously proposed robust Capon beamformers in the presence of large DOA mismatch and other array imperfections.

Keywords: passive sonar; weak target detection; robust Capon beamforming; large DOA mismatch; two-step steering vector estimation

---

1. Introduction

Underwater target detection is a primordial task for passive sonar systems. In practical underwater environments, the presence of the strong underwater targets will severely affect the detection performance of the weak targets. Therefore, the strong targets are considered as interferences for the weak targets. Capon beamforming [1,2], as an important pre-processing technique in array signal processing, is often applied in passive sonar arrays for improving the detectability of weak underwater targets [3–5]. By extracting the signal of weak targets and suppressing the strong interferences and noise at the beamformer output simultaneously, Capon beamformer can obtain the optimal output signal-to-interference-plus-noise ratio (SINR) and prevent the weak sources from submerging in strong interferences. However, due to some imperfections in practice, such as direction-of-arrival (DOA) mismatch, imperfect array element calibration, and distorted array shape [6–8], the knowledge of the actual steering vector can be inaccurate, which will cause a substantial degradation in the performance of Capon beamforming.

In recent years, various approaches have been developed to improve the robustness of Capon beamformer against the steering vector error. Generally, these approaches can be classified into two categories: The robust constrained approach and the steering vector estimation approach. In the robust constrained approach, such as the worst-case performance optimization-based approach [9] and the doubly constrained approach [10,11], the uncertainty constraints on the array magnitude response or the steering vector error are usually imposed without estimating the actual steering vector. However, in the case of large steering vector error, a large size of uncertainty set is required in the robust constrained approach which will weaken the abilities of interference-plus-noise suppression. On the other hand, the main-lobe of robust constrained beamformer usually points to the presumed DOA of the signal rather than the actual DOA. Unlike the robust constrained approach, the steering vector estimation approach searches for the actual steering vector directly based on some prior information. The most typical beamformer based on steering vector estimation is the so-called eigenspace-based beamformer (ESB) [12], where the actual steering vector can be well estimated by utilizing the signal-plus-interference subspace. The ESB is considered to be one of the most powerful techniques robust to arbitrary steering vector mismatch case [9]. Nevertheless, it will suffer from severe performance degradation if the dimension of the signal-plus-interference subspace is misestimated. To solve this problem, some improved steering vector estimation approaches were proposed. In Reference [13], the actual steering vector is estimated iteratively where a quadratic convex optimization problem is solved at each iteration. The main drawback of this method is the high computational complexity due to the iterations. In Reference [14], a nonconvex quadratically constrained quadratic programming (QCQP) problem and the corresponding relaxation procedure were developed to estimate the actual steering vector. As compared with the beamformer in Reference [13], this method has more degrees of freedom and achieves superior performance.

This paper mainly focuses on improving the robustness of Capon beamformer against the steering vector error dominated by large DOA mismatch. This type of steering vector error is very common in the practical applications of passive sonar, where the accurate prior information of the DOA of the signal is usually unavailable [15]. In this case, large DOA mismatch is naturally dominant in the steering vector error, but other array imperfections can also occur. Besides the beamformers discussed above, some robust Capon beamformers specially developed to improve the robustness against the steering vector error dominated by large DOA mismatch have been proposed. Generally, these beamformers belong to the steering vector estimation approach. In Reference [16], the actual steering vector is considered to lie in the intersection of two constructed subspaces and can be estimated by the alternating projection algorithm. However, similar to the ESB, an incorrect estimation of the dimension of signal-plus-interference subspace will cause performance degradation. In Reference [17], a robust Capon beamforming method against large DOA mismatch was proposed. This method expresses the actual steering vector as a linear combination of the columns of a subspace and estimate the coefficients by maximizing the output power. However, this beamformer lacks robustness against other array imperfections. To solve this, an improved beamformer was proposed in Reference [18]. By employing the general-rank signal model-based robust beamforming technique [19], this beamformer provides further robustness against other array imperfections.

In this paper, we develop a new approach to robust Capon beamforming in the presence of large DOA mismatch and other array imperfections. Our approach is based on the two-step estimation of the actual steering vector. By projecting the actual steering vector obliquely onto a subspace, we can decompose the actual steering vector into two components and then search for them within the subspace and the neighborhood of the oblique projection steering vector, respectively. It turns out the natural formulations of the two estimation problems all involve nonconvex objective function or nonconvex constraint, and therefore are NP-hard to solve. To mitigate this problem, we utilize the semidefinite relaxation technique [20] and Charnes-Cooper transformation technique [21] to derive convex formulations of two estimation problems, and obtain the optimal solutions by applying the rank-one decomposition theorem [22]. Numerical simulations show that the performance of the proposed beamformer is always close to the optimal value in the presence of large DOA mismatch and other array imperfections.

This paper is organized as follows. In Section 2, the array signal model and some backgrounds on Capon beamforming are introduced. In Section 3, the new robust Capon beamforming method is proposed. In Section 4, simulation results are presented to demonstrate the effectiveness of the proposed method. Finally, the paper is concluded in Section 5.

2. Backgrounds

2.1. Array Signal Model

Consider a linear sonar array with M omnidirectional elements receiving narrowband signals from $D(D < M)$ far-field underwater targets (one desired signal and $D-1$ interferences). The depiction on the array signal model is shown in Figure 1. The kth array snapshot vector $\mathbf{x}(k) \in \mathbb{C}^M$ can be modeled as:

$$\mathbf{x}(k) = \mathbf{x}_s(k) + \mathbf{x}_{\text{int}}(k) + \mathbf{n}(k) \tag{1}$$

where $\mathbf{x}_s(k)$, $\mathbf{x}_{\text{int}}(k)$, and $\mathbf{n}(k)$ denote the signal of interest (SOI), interference and noise components, respectively. The SOI can be written as $\mathbf{x}_s(k) = \mathbf{a}s_0(k)$, where $\mathbf{a}$ is the steering vector associated with the SOI and $s_0(k)$ is the waveform of the SOI. The interferences can be written as $\mathbf{x}_{\text{int}}(k) = \sum_{i=1}^{D-1} \mathbf{a}_i s_i(k)$, where $\mathbf{a}_i$ and $s_i(k)$ denote the steering vector and the waveform of the ith interference, respectively. The additive noise $\mathbf{n}(k)$ is modeled as a zero-mean spatially and temporally white Gaussian process with the power σ_n^2. Here, the SOI, interferences, and noise are statistically independent with each other. As shown in Figure 1, we define an angular region of interest (ROI) Θ and its complementary region $\overline{\Theta}$, that is, $\Theta \cup \overline{\Theta}$ covers the whole spatial domain and $\Theta \cap \overline{\Theta}$ is empty. The main required prior information is that the SOI is located in Θ while the interferences are not, which implies that the accurate DOA of the SOI is not required. Consequently, the DOA uncertainty set is Θ which leads to a large DOA mismatch scenario.

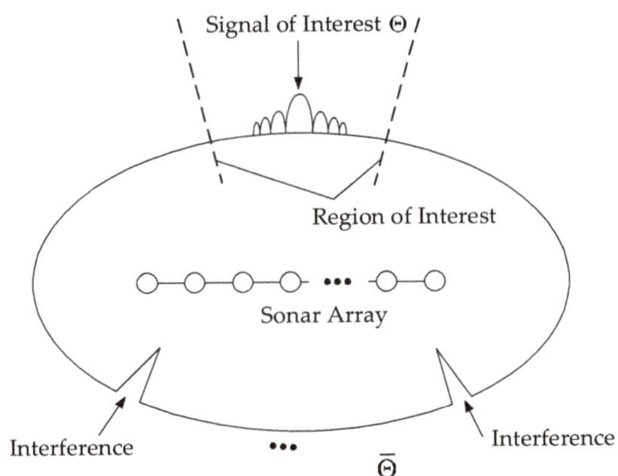

Figure 1. Depiction on array signal model.

2.2. Capon Beamforming

The output of a beamformer is given by:

$$y(k) = \mathbf{w}^H \mathbf{x}(k) \tag{2}$$

where $\mathbf{w} = [w_1, w_2, \cdots, w_M]^T \in \mathbb{C}^M$ is the complex weight vector, $(\cdot)^T$ and $(\cdot)^H$ denote the transpose and Hermitian transpose, respectively. The output SINR of a beamformer is defined as:

$$\text{SINR}_{\text{out}} = \frac{E\left\{|\mathbf{w}^H \mathbf{x}_s(k)|^2\right\}}{E\left\{|\mathbf{w}^H(\mathbf{x}_{\text{int}}(k) + \mathbf{n}(k))|^2\right\}} = \frac{\sigma_0^2|\mathbf{w}^H\mathbf{a}|^2}{\mathbf{w}^H\mathbf{R}_{i+n}\mathbf{w}} \tag{3}$$

where $\sigma_0^2 = E\left\{|s_0(k)|^2\right\}$ is the power of the SOI, $\mathbf{R}_{i+n} = E\left\{(\mathbf{x}_{\text{int}}(k) + \mathbf{n}(k))(\mathbf{x}_{\text{int}}(k) + \mathbf{n}(k))^H\right\}$ is the interference-plus-noise covariance matrix and $E\{\cdot\}$ denotes the statistical expectation. The Capon beamformer aims at maximizing the output SINR by maintaining a distortionless response towards the SOI and minimizing the output power of interference-plus-noise. Therefore, the Capon beamformer is equivalent to:

$$\min_{\mathbf{w}}\mathbf{w}^H\mathbf{R}_{i+n}\mathbf{w} \qquad \text{subject to } \mathbf{w}^H\mathbf{a} = 1 \tag{4}$$

Based on the statistical independence among the SOI, interferences, and noise, we can easily obtain that:

$$\mathbf{R}_x = E\left\{\mathbf{x}(k)\mathbf{x}^H(k)\right\} = \sigma_0^2\mathbf{a}\mathbf{a}^H + \mathbf{R}_{i+n} \tag{5}$$

$$\mathbf{w}^H\mathbf{R}_x\mathbf{w} = \mathbf{w}^H\mathbf{R}_{i+n}\mathbf{w} + \sigma_0^2\left|\mathbf{w}^H\mathbf{a}\right|^2 \tag{6}$$

Hence, Problem (4) is equivalent to:

$$\min_{\mathbf{w}}\mathbf{w}^H\mathbf{R}_x\mathbf{w} \qquad \text{subject to } \mathbf{w}^H\mathbf{a} = 1 \tag{7}$$

The optimal solution to Problem (7) yields to the weight of standard Capon beamformer (SCB) [2]:

$$\mathbf{w} = \frac{\mathbf{R}_x^{-1}\mathbf{a}}{\mathbf{a}^H\mathbf{R}_x^{-1}\mathbf{a}} \tag{8}$$

In practical applications, $\mathbf{R}_x$ is usually unavailable and replaced by the sample covariance matrix:

$$\hat{\mathbf{R}}_x = \frac{1}{N}\sum_{k=1}^{N} x(k)x^H(k) \tag{9}$$

where N is the number of snapshots.

2.3. Effects of Steering Vector Error

It is worth noting that the weight of SCB is based on the assumption that the steering vector of the SOI is accurately known. In practice, factors such as DOA mismatch and array imperfections (including array element calibration errors and position errors) can lead to an error between the presumed steering vector and the actual steering vector, i.e.:

$$\mathbf{a} = \bar{\mathbf{a}} + \mathbf{e} \tag{10}$$

where $\bar{\mathbf{a}}$ is the presumed steering vector and $\mathbf{e}$ is an unknown error vector. The weight vector of SCB, when using the sample covariance matrix $\hat{\mathbf{R}}_x$ and the presumed steering vector $\bar{\mathbf{a}}$, is then given by:

$$\mathbf{w} = \frac{\hat{\mathbf{R}}_x^{-1}\bar{\mathbf{a}}}{\bar{\mathbf{a}}^H\hat{\mathbf{R}}_x^{-1}\bar{\mathbf{a}}} \tag{11}$$

With Problem (11), the array response at the direction of the SOI can be expressed as follows:

$$H(\mathbf{a}) = \left| \mathbf{w}^H \mathbf{a} \right| = \frac{\left| \overline{\mathbf{a}}^H \hat{\mathbf{R}}_x^{-1} \mathbf{a} \right|}{\left| \overline{\mathbf{a}}^H \hat{\mathbf{R}}_x^{-1} \overline{\mathbf{a}} \right|} = \frac{\left| \overline{\mathbf{a}}^H \mathbf{V}\mathbf{V}^H \mathbf{a} \right|}{\left| \overline{\mathbf{a}}^H \mathbf{V}\mathbf{V}^H \overline{\mathbf{a}} \right|} \tag{12}$$

where $\mathbf{V} \triangleq \hat{\mathbf{R}}_x^{-1/2}$. Applying the Cauchy-Schwarz inequality, we have that:

$$\left| \overline{\mathbf{a}}^H \mathbf{V}\mathbf{V}^H \mathbf{a} \right| \leq \|\mathbf{a}^H \mathbf{V}\| \|\overline{\mathbf{a}}^H \mathbf{V}\| \tag{13}$$

where $\|\cdot\|$ denotes the 2-norm operation. By substituting Problem (13) into Problem (12), we can write that:

$$H(\mathbf{a}) \leq \frac{\|\mathbf{a}^H \mathbf{V}\| \|\overline{\mathbf{a}}^H \mathbf{V}\|}{\|\overline{\mathbf{a}}^H \mathbf{V}\|^2} = \frac{\|\mathbf{a}^H \mathbf{V}\|}{\|\overline{\mathbf{a}}^H \mathbf{V}\|} = \frac{\sqrt{1/\overline{\mathbf{a}}^H \hat{\mathbf{R}}_x^{-1} \overline{\mathbf{a}}}}{\sqrt{1/\mathbf{a}^H \hat{\mathbf{R}}_x^{-1} \mathbf{a}}} \tag{14}$$

Based on the Capon spatial spectrum estimator [1], $1/\mathbf{a}^H \hat{\mathbf{R}}_x^{-1} \mathbf{a}$ and $1/\overline{\mathbf{a}}^H \hat{\mathbf{R}}_x^{-1} \overline{\mathbf{a}}$ can be considered as the power collected from the directions of the SOI and the presumed steering vector, respectively. Also, it is easy to know that $1/\mathbf{a}^H \hat{\mathbf{R}}_x^{-1} \mathbf{a}$ can be approximated as the power of SOI-plus-noise. Since the Capon estimator has high resolution in spectrum estimation, $1/\mathbf{a}^H \hat{\mathbf{R}}_x^{-1} \mathbf{a}$ will deviate from $1/\overline{\mathbf{a}}^H \hat{\mathbf{R}}_x^{-1} \overline{\mathbf{a}}$ as long as the steering vector error occurs between $\mathbf{a}$ and $\overline{\mathbf{a}}$, resulting in the distortion of the array response. Particularly when large steering vector error occurs, $1/\overline{\mathbf{a}}^H \hat{\mathbf{R}}_x^{-1} \overline{\mathbf{a}}$ can be approximated as the power of noise only. In this case, the value of $H(\mathbf{a})$ will decrease as the signal-to-noise ratio (SNR) increases, which means the enhancement of the distortion at the direction of the SOI. The enhanced distortion can cause the signal self-nulling phenomenon [13].

As we discussed in Section 1, the steering vector error caused by large DOA mismatch is dominant in the applications of passive sonar systems. Motivated by this fact, this paper mainly focuses on estimating the actual steering vector to improve the robustness of Capon beamformer in the case of large DOA mismatch, but unlike the beamformer in Reference [17], other array imperfections are also taken into consideration.

3. Proposed Method

Our approach is based on the two-step steering vector estimation. We begin with the decomposition of the actual steering vector and then formulate two optimization problems to estimate the actual steering vector in two steps. The corrected beam can be formed by using the estimated actual steering vector.

3.1. Actual Steering Vector Decomposition

Theoretically, we can build a positive definite matrix $\widetilde{\mathbf{C}} = \int_{\Theta} \mathbf{a}(\theta)\mathbf{a}^H(\theta)d\theta$ and obtain a subspace $\widetilde{\mathbf{U}}$ spanned by the principal eigenvectors $\{\tilde{\mathbf{u}}_m\}_{m=1}^{L}$ of $\widetilde{\mathbf{C}}$ corresponding to the L largest eigenvalues:

$$\widetilde{\mathbf{U}} = [\tilde{\mathbf{u}}_1, \tilde{\mathbf{u}}_2, \cdots, \tilde{\mathbf{u}}_L] \tag{15}$$

Here, $\mathbf{a}(\theta)$ denotes the actual steering vector whose DOA is Θ. Since the L largest eigenvalues contain most of the energy of the eigenvalues, any actual steering vector $\{\mathbf{a}(\theta)|\theta \in \Theta\}$ is believed to lie within the subspace $\widetilde{\mathbf{U}}$. Hence, the actual steering vector can be expressed as a linear combination of the columns of $\widetilde{\mathbf{U}}$:

$$\mathbf{a} = \widetilde{\mathbf{U}}\mathbf{r} \tag{16}$$

where $\mathbf{r}$ is the coefficient vector. In fact, $\widetilde{\mathbf{U}}$ is unavailable in practice due to the imperfect knowledge of the actual steering vector. Instead, we build another subspace $\mathbf{U}$ spanned by the set of presumed steering vectors $\left\{\overline{\mathbf{a}}(\theta)\middle|\theta\in\Theta\right\}$:

$$\mathbf{U} = [\mathbf{u}_1, \mathbf{u}_2, \cdots, \mathbf{u}_L] \tag{17}$$

where $\{\widetilde{\mathbf{u}}_m\}_{m=1}^{L}$ denotes the L principal eigenvectors of $\mathbf{C} = \int_{\Theta}\overline{a}(\theta)\overline{a}^H(\theta)d\theta$. It can be easily seen that the matrix $\mathbf{C}$ will deviate from the matrix $\widetilde{\mathbf{C}}$ when steering vector errors occur, leading to a mismatch between the subspaces $\widetilde{\mathbf{U}}$ and $\mathbf{U}$:

$$\widetilde{\mathbf{U}} = \mathbf{U} + \Delta_{\mathbf{U}} \tag{18}$$

where $\Delta_{\mathbf{U}}$ is the $M \times L$ mismatch matrix. By substituting Problem (18) into Problem (16), $\mathbf{a}$ can be decomposed into two components, i.e.:

$$\mathbf{a} = \mathbf{a}_{\mathbf{U}} + \mathbf{e}_{\Delta} \tag{19}$$

Here, we define $\mathbf{a}_{\mathbf{U}} \triangleq \mathbf{U}\mathbf{r}$ and $\mathbf{e}_{\Delta} \triangleq \Delta_{\mathbf{U}}\mathbf{r}$. It can be seen that $\mathbf{a}_{\mathbf{U}}$ is a vector lying within the subspace $\mathbf{U}$ and $\mathbf{e}_{\Delta}$ is an error vector caused by $\Delta_{\mathbf{U}}$. It is worth mentioning that in the case of only DOA mismatch, we have $\mathbf{U} = \widetilde{\mathbf{U}}$, as long as Θ contains the actual DOA of the SOI and avoids all the interferences. In other words, the existence of DOA mismatch will not result in the mismatch between $\mathbf{U}$ and $\widetilde{\mathbf{U}}$. Consequently, the error vector $\mathbf{e}_{\mathbf{U}} \triangleq \mathbf{a}_{\mathbf{U}} - \mathbf{a}$, which also lies within $\mathbf{U}$, is dominated by DOA mismatch while the error vector $\mathbf{e}_{\Delta}$ is dominated by other array imperfections. Here, the presumed steering vector $\overline{\mathbf{a}}$ is an arbitrary element in the set of $\left\{\overline{\mathbf{a}}(\theta)\middle|\theta\in\Theta\right\}$.

The relationship among $\widetilde{\mathbf{U}}$, $\mathbf{U}$, $\mathbf{e}_{\mathbf{U}}$, $\mathbf{e}_{\Delta}$, $\overline{\mathbf{a}}$, $\mathbf{a}_{\mathbf{U}}$, and $\mathbf{a}$ is depicted geometrically in Figure 2. As shown, $\mathbf{a}_{\mathbf{U}}$ can be considered as the oblique projection vector of the actual steering vector $\mathbf{a}$ onto the subspace $\mathbf{U}$ and the projection direction is opposite to the direction of the error vector $\mathbf{e}_{\Delta}$. The basic idea behind our approach is to estimate the actual steering vector $\mathbf{a}$ in two steps. We first search for the oblique projection steering vector $\mathbf{a}_{\mathbf{U}}$ within the subspace $\mathbf{U}$ and then search for the actual steering vector $\mathbf{a}$ within the neighborhood of $\mathbf{a}_{\mathbf{U}}$. In the following subsections, different optimization principles are utilized to estimate the oblique projection steering vector and the actual steering vector, respectively.

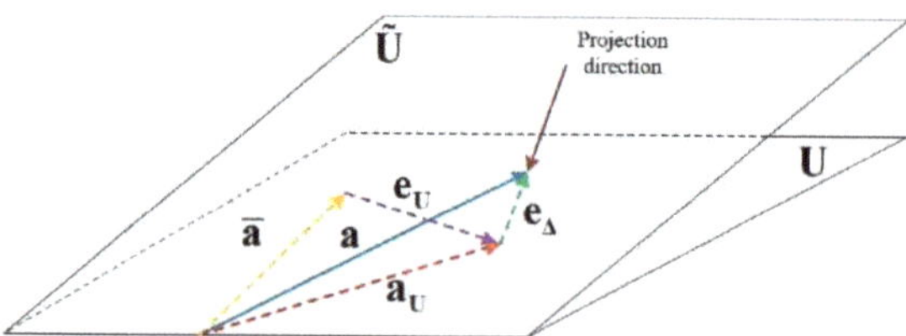

Figure 2. Geometrical depiction on the relationship among the subspaces $\widetilde{\mathbf{U}}$ and $\mathbf{U}$, the error vectors $\mathbf{e}_{\mathbf{U}}$ and $\mathbf{e}_{\Delta}$, the steering vector $\overline{\mathbf{a}}$, $\mathbf{a}_{\mathbf{U}}$, and $\mathbf{a}$.

3.2. Oblique Projection Steering Vector Estimation

Since $\mathbf{a}_{\mathbf{U}}$ lies within the subspace $\mathbf{U}$, it can be expressed as the linear combination of the columns of $\mathbf{U}$:

$$\mathbf{a}_{\mathbf{U}} = \alpha_1\mathbf{u}_1 + \alpha_2\mathbf{u}_2 + \cdots + \alpha_L\mathbf{u}_L \tag{20}$$

By multiplying $\mathbf{u}_k^H(k = 1, \cdots, L)$ on both sides of (20), we can easily obtain the coefficients $\alpha_k = \mathbf{u}_k^H \mathbf{a_U}(k = 1, 2, \cdots, L)$. Then, Problem (20) can be rewritten as:

$$\mathbf{a_U} = \mathbf{U}\mathbf{U}^H \mathbf{a_U} \qquad (21)$$

By making use of Problem (21), the oblique projection $\mathbf{a_U}$ can be forced to belong to the subspace $\mathbf{U}$. Additionally, an equality constraint $\|\mathbf{a_U}\| = \sqrt{M}$, which maintains the norm of the estimated $\mathbf{a_U}$, is imposed to avoid the scaling ambiguity. Now we can search for $\mathbf{a_U}$ by maximizing the output power $1/\mathbf{a_U}^H \hat{\mathbf{R}}_x^{-1} \mathbf{a_U}$, which can be formulated as the following optimization problem:

$$\begin{aligned} \min_{\mathbf{a_U}} \quad & \mathbf{a_U}^H \hat{\mathbf{R}}_x^{-1} \mathbf{a_U} \\ \text{subject to} \quad & \mathbf{a_U} = \mathbf{U}\mathbf{U}^H \mathbf{a_U} \\ & \|\mathbf{a_U}\| = \sqrt{M} \end{aligned} \qquad (22)$$

By substituting the equality constraint $\mathbf{a_U} = \mathbf{U}\mathbf{U}^H \mathbf{a_U}$ into the objective function, Problem (22) can be modified as:

$$\begin{aligned} \min_{\mathbf{a_U}} \quad & \mathbf{a_U}^H \mathbf{U}\mathbf{U}^H \hat{\mathbf{R}}_x^{-1} \mathbf{U}\mathbf{U}^H \mathbf{a_U} \\ \text{subject to} \quad & \|\mathbf{a_U}\| = \sqrt{M} \end{aligned} \qquad (23)$$

It is worth noting that the solution to Problem (23) may result in the magnification of output noise especially at low SNR, thus an extra constraint $\left| \mathbf{a_U}^H \bar{\mathbf{a}}(\theta) \right| \leq \left| \bar{\mathbf{a}}^{-H-}\mathbf{a}(\theta) \right|$ for any $\theta \in \overline{\Theta}$ is invoked to control the power from the sidelobe region $\overline{\Theta}$. This constraint can be equivalently expressed as $\mathbf{a_U}^H \mathbf{C}\mathbf{a_U} \leq \bar{\mathbf{a}}^{-H-} \mathbf{C}\bar{\mathbf{a}}$, where $\mathbf{C} = \int_{\overline{\Theta}} \bar{\mathbf{a}}(\theta)\bar{\mathbf{a}}^H(\theta)d\theta$ [13]. Now, the estimation problem can be written as:

$$\begin{aligned} \min_{\mathbf{a_U}} \quad & \mathbf{a_U}^H \mathbf{U}\mathbf{U}^H \hat{\mathbf{R}}_x^{-1} \mathbf{U}\mathbf{U}^H \mathbf{a_U} \\ \text{subject to} \quad & \mathbf{a_U}^H \mathbf{C}\mathbf{a_U} \leq \bar{\mathbf{a}}^{-H-} \mathbf{C}\bar{\mathbf{a}} \\ & \|\mathbf{a_U}\| = \sqrt{M} \end{aligned} \qquad (24)$$

It is interesting that the steering vector estimation problem in Reference [17] has some insightful connections to Problem (24). Since the approach in Reference [17] only considers the factor of DOA mismatch, the actual steering vector is represented as $\mathbf{a} = \mathbf{Ur}$ directly, where $\mathbf{r}$ denotes the coefficient vector. The origin estimation problem in Reference [17] can be expressed in terms of $\mathbf{r}$ as:

$$\begin{aligned} \min_{\mathbf{r}} \quad & \mathbf{r}^H \mathbf{U}^H \hat{\mathbf{R}}_x^{-1} \mathbf{Ur} \\ \text{subject to} \quad & \|\mathbf{r}\| = \sqrt{M} \end{aligned} \qquad (25)$$

Using Problem (21), we can easily obtain $\mathbf{r} = \mathbf{U}^H \mathbf{a}$ and then Problem (25) can be equivalently rewritten as (23). As we discussed above, an extra sidelobe control constraint is introduced in Problem (24) as compared with Problem (23). Consequently, it is expected that the proposed method can provide better performance than the beamformer in Reference [17] in the case of large DOA mismatch, especially in the low SNR region.

Unfortunately, the QCQP Problem (24) is a nonconvex NP-hard problem because of the equality constraint. Next, we focus on solving Problem (24) in polynomial time. Using the equalities

$\mathbf{a}_U^H \mathbf{U}\mathbf{U}^H \hat{\mathbf{R}}_x^{-1} \mathbf{U}\mathbf{U}^H \mathbf{a}_U = tr(\mathbf{U}\mathbf{U}^H \hat{\mathbf{R}}_x^{-1} \mathbf{U}\mathbf{U}^H \mathbf{a}_U \mathbf{a}_U^H)$, $\mathbf{a}_U^H \bar{\mathbf{C}} \mathbf{a}_U = tr(\bar{\mathbf{C}} \mathbf{a}_U \mathbf{a}_U^H)$, and $\mathbf{a}_U^H \mathbf{a}_U = tr(\mathbf{a}_U \mathbf{a}_U^H)$, where $tr(\cdot)$ represents the trace of a matrix, Problem (24) can be equivalently written as:

$$\min_{\mathbf{a}_U} tr(\mathbf{U}\mathbf{U}^H \hat{\mathbf{R}}_x^{-1} \mathbf{U}\mathbf{U}^H \mathbf{a}_U \mathbf{a}_U^H)$$
$$\text{subject to } tr(\bar{\mathbf{C}} \mathbf{a}_U \mathbf{a}_U^H) \leq \bar{\mathbf{a}}^H \bar{\mathbf{C}} \bar{\mathbf{a}} \tag{26}$$
$$tr(\mathbf{a}_U \mathbf{a}_U^H) = M$$

By introducing the positive semidefinite matrix variable $\mathbf{Q} \triangleq \mathbf{a}_U \mathbf{a}_U^H$ and using the semidefinite relaxation technique [20], we obtain the following relaxed problem:

$$\min_{\mathbf{Q}} tr(\mathbf{R}_0 \mathbf{Q})$$
$$\text{subject to } tr(\mathbf{R}_1 \mathbf{Q}) \leq \beta \tag{27}$$
$$tr(\mathbf{R}_2 \mathbf{Q}) = M$$
$$\mathbf{Q} \geq 0$$

where $\mathbf{R}_0 \triangleq \mathbf{U}\mathbf{U}^H \hat{\mathbf{R}}_x^{-1} \mathbf{U}\mathbf{U}^H$, $\mathbf{R}_1 \triangleq \bar{\mathbf{C}}$, $\mathbf{R}_2 \triangleq \mathbf{I}_M$, $\beta \triangleq \bar{\mathbf{a}}^H \bar{\mathbf{C}} \bar{\mathbf{a}}$. Note that we drop the nonconvex rank-one constraint on $\mathbf{Q}$ and just require $\mathbf{Q} \geq 0$. Problem (27) is a convex semidefinite program (SDP) problem and can be easily solved. Generally, the solution to the SDP problem may not be exactly rank-one. Thus, we first check the rank of the optimal solution $\mathbf{Q}^*$. If $rank(\mathbf{Q}^*) = 1$, then we can easily write $\mathbf{Q}^* = \mathbf{a}_U^* \mathbf{a}_U^{*H}$ by eigen-decomposition and $\mathbf{a}_U^*$ will be the optimal solution to (24). If $rank(\mathbf{Q}^*) \geq 2$, then we can adopt the rank-one decomposition theorem [22] to find a rank-one optimal solution $\mathbf{a}_U^* \mathbf{a}_U^{*H}$ to Problem (27), and $\mathbf{a}_U^*$ is optimal for Problem (24). The rank-one decomposition theorem [22] is cited as the following lemma.

Lemma 1. *Let* $\mathbf{A}_i$ *($i = 0, 1, 2$) be* $M \times M$ *($M \geq 3$) complex Hermitian matrices, and* $\mathbf{X}$ *be a nonzero Hermitian positive semidefinite matrix with rank r.*

If $r \geq 3$, *then one can find a nonzero vector* $\mathbf{y} \in \text{Range}(\mathbf{X})$ *(synthetically denoted as* $\mathbf{y} = D_1(\mathbf{X}, \mathbf{A}_0, \mathbf{A}_1, \mathbf{A}_2)$*), such that:*

$$tr(\mathbf{A}_i \mathbf{y}\mathbf{y}^H) = tr(\mathbf{A}_i \mathbf{X}), \ i = 0, 1, 2 \tag{28}$$

If $r = 2$, *then for any* $\mathbf{z} \notin \text{Range}(\mathbf{X})$, *one can find a nonzero vector* $\mathbf{y} \in \mathbb{C}^M$ *in the linear subspace spanned by* $\mathbf{z}$ *and* $\text{Range}(\mathbf{X})$ *(synthetically denoted as* $\mathbf{y} = D_2(\mathbf{X}, \mathbf{A}_0, \mathbf{A}_1, \mathbf{A}_2)$*), such that:*

$$tr(\mathbf{A}_i \mathbf{y}\mathbf{y}^H) = tr(\mathbf{A}_i \mathbf{X}), \ i = 0, 1, 2 \tag{29}$$

It follows from Lemma 1 that if $rank(\mathbf{Q}^*) \geq 2$, we can find a vector $\mathbf{a}_U^*$ such that:

$$tr(\mathbf{R}_i \mathbf{a}_U^* \mathbf{a}_U^{*H}) = tr(\mathbf{R}_i \mathbf{Q}^*), \ i = 0, 1, 2 \tag{30}$$

Hence, $\mathbf{a}_U^* \mathbf{a}_U^{*H}$ satisfies all the constraints in Problem (27) and the objective function evaluated at $\mathbf{a}_U^*$ is equal to the optimal value of Problem (27); as a consequence, $\mathbf{a}_U^* \mathbf{a}_U^{*H}$ is optimal for Problem (27) and $\mathbf{a}_U^*$ is an optimal solution to Problem (24). The procedure to find the optimal solution $\mathbf{a}_U^*$ is summarized in Algorithm 1.

Algorithm 1. Procedure leading to an optimal solution to (24)

Input: $\hat{\mathbf{R}}_x$, $\bar{\mathbf{a}}$, $\mathbf{U}$ and $\mathbf{C}$
Output: An optimal solution $\mathbf{a}_U^*$ to (24)
1: solve the SDP problem (27) and find an optimal solution $\mathbf{Q}^*$.
2: **if** $rank(\mathbf{Q}^*) = 1$, **then**
3: find the optimal solution $\mathbf{a}_U^*$ via eigen-decomposition, i.e., $\mathbf{Q}^* = \mathbf{a}_U^* \mathbf{a}_U^{*H}$
4: **else if** $rank(\mathbf{Q}^*) = 2$, **then**
5: find $\mathbf{a}_U^* = D_2(\mathbf{Q}^*, \mathbf{R}_0, \mathbf{R}_1, \mathbf{R}_2)$
6: **else, then**
7: find $\mathbf{a}_U^* = D_1(\mathbf{Q}^*, \mathbf{R}_0, \mathbf{R}_1, \mathbf{R}_2)$
8: **end if**
9: **return** $\mathbf{a}_U^*$

3.3. Actual Steering Vector Estimation

Besides the DOA mismatch, the Capon beamformer is also very sensitive to other array imperfections, which contributes to the error vector $\mathbf{e}_\Delta$ as shown in Figure 2. Based on this fact, we impose a bounded-norm constraint on the error vector $\mathbf{e}_\Delta$ and assume that the actual steering vector $\mathbf{a}$ lies in a spherical uncertainty set $S_\mathbf{a}$:

$$\mathbf{a} \in S_\mathbf{a} \triangleq \{\mathbf{a} | \|\mathbf{a} - \mathbf{a}_U\| \leq \gamma\} \tag{31}$$

where $\gamma > 0$ is a given constant. In the following derivations, we design a novel objective function to search for the actual steering vector in $S_\mathbf{a}$. Essentially, the novel objective function is based on maximizing the output SINR. Based on the introduced array signal model, the output SINR (3) can be reformulated as follows:

$$\begin{aligned}
\text{SINR}_{\text{out}} &= \frac{\mathbf{w}^H \mathbf{R}_s \mathbf{w}}{\mathbf{w}^H \mathbf{R}_{\text{int}} \mathbf{w} + \mathbf{w}^H \mathbf{R}_n \mathbf{w}} \\
&= \frac{\sigma_0^2 |\mathbf{w}^H \mathbf{a}|^2}{\sum\limits_{i=1}^{D-1} \sigma_i^2 |w^H a_i|^2 + \sigma_n^2 w^H w}
\end{aligned} \tag{32}$$

Utilizing the weight of the SCB (8), the array response of the SOI, i.e., $|\mathbf{w}^H \mathbf{a}|$, is 1, and the array response of the interference can be obtained by:

$$\begin{aligned}
H(\mathbf{a}_i) &= |\mathbf{w}^H \mathbf{a}_i| \\
&= \frac{|\mathbf{a}^H \mathbf{R}_x^{-1} \mathbf{a}_i|}{|\mathbf{a}^H \mathbf{R}_x^{-1} \mathbf{a}|} = \frac{|\mathbf{a}^H \mathbf{T} \mathbf{T}^H \mathbf{a}_i|}{|\mathbf{a}^H \mathbf{T} \mathbf{T}^H \mathbf{a}|}
\end{aligned} \tag{33}$$

where $\mathbf{T} \triangleq \mathbf{R}_x^{-1/2}$. Applying the Cauchy-Schwarz inequality $|\mathbf{a}^H \mathbf{T} \mathbf{T}^H \mathbf{a}_i| \leq \|\mathbf{a}^H \mathbf{T}\| \|\mathbf{a}_i^H \mathbf{T}\|$, we can easily have that:

$$H(\mathbf{a}_i) \leq \frac{\|\mathbf{a}^H \mathbf{T}\| \|\mathbf{a}_i^H \mathbf{T}\|}{\|\mathbf{a}^H \mathbf{T}\|^2} = \frac{\sqrt{1/\mathbf{a}^H \mathbf{R}_x^{-1} \mathbf{a}}}{\sqrt{1/\mathbf{a}_i^H \mathbf{R}_x^{-1} \mathbf{a}_i}} \tag{34}$$

Moreover, the power of the SOI and the interferences can be given by:

$$\sigma_0^2 = \frac{1}{\mathbf{a}^H \mathbf{R}_x^{-1} \mathbf{a}}, \quad \sigma_i^2 = \frac{1}{\mathbf{a}_i^H \mathbf{R}_x^{-1} \mathbf{a}_i} \tag{35}$$

By substituting Problem (33)–(35) into Problem (32), we can write that:

$$
\begin{aligned}
\text{SINR}_{out} \;&\geq\; \frac{\dfrac{1}{\mathbf{a}^H\mathbf{R}_x^{-1}\mathbf{a}}}{\dfrac{D-1}{\mathbf{a}^H\mathbf{R}_x^{-1}\mathbf{a}}+\sigma_{\tilde{n}}^2\,\dfrac{\mathbf{a}^H\mathbf{R}_x^{-2}\mathbf{a}}{\left(\mathbf{a}^H\mathbf{R}_x^{-1}\mathbf{a}\right)^2}} \\[2mm]
&=\; \frac{1}{D-1+\sigma_{\tilde{n}}^2\,\dfrac{\mathbf{a}^H\mathbf{R}_x^{-2}\mathbf{a}}{\mathbf{a}^H\mathbf{R}_x^{-1}\mathbf{a}}}
\end{aligned}
\tag{36}
$$

Hence, the maximization of the output SINR is equivalent to:

$$
\max_{\mathbf{a}}\frac{\mathbf{a}^H\mathbf{R}_x^{-1}\mathbf{a}}{\mathbf{a}^H\mathbf{R}_x^{-2}\mathbf{a}}
\tag{37}
$$

Using the sample covariance matrix $\hat{\mathbf{R}}_x^{-1}$, the following optimization problem is constructed to search for the actual steering vector of the SOI in the uncertainty set:

$$
\max_{\mathbf{a}}\frac{\mathbf{a}^H\hat{\mathbf{R}}_x^{-1}\mathbf{a}}{\mathbf{a}^H\hat{\mathbf{R}}_x^{-2}\mathbf{a}}
\tag{38}
$$
$$
\text{subject to } \|\mathbf{a}-\mathbf{a}_U\|\leq\gamma
$$

Problem (38) is a nonconvex fractional quadratic optimization problem. With an extra variable t ($t^2=1$), it can be transformed into a homogeneous form:

$$
\max_{\mathbf{g}}\frac{tr(\mathbf{P}_0\mathbf{g}\mathbf{g}^H)}{tr(\mathbf{P}_1\mathbf{g}\mathbf{g}^H)}
\tag{39}
$$
$$
\text{subject to } tr(\mathbf{P}_2\mathbf{g}\mathbf{g}^H)\leq\gamma^2
$$
$$
tr(\mathbf{P}_3\mathbf{g}\mathbf{g}^H)=1
$$

where $\mathbf{g}\triangleq\begin{bmatrix}\mathbf{a}\\ t\end{bmatrix}$, $\mathbf{P}_0\triangleq\begin{bmatrix}\hat{\mathbf{R}}_x^{-1}&\mathbf{0}_{M\times1}\\ \mathbf{0}_{1\times M}&0\end{bmatrix}$, $\mathbf{P}_1\triangleq\begin{bmatrix}\hat{\mathbf{R}}_x^{-2}&\mathbf{0}_{M\times1}\\ \mathbf{0}_{1\times M}&0\end{bmatrix}$, $\mathbf{P}_2\triangleq\begin{bmatrix}\mathbf{I}_M&-\mathbf{a}_U\\ -\mathbf{a}_U^H&\mathbf{a}_U^H\mathbf{a}_U\end{bmatrix}$, $\mathbf{P}_3\triangleq$ $\begin{bmatrix}\mathbf{0}_{M\times M}&\mathbf{0}_{M\times1}\\ \mathbf{0}_{1\times M}&1\end{bmatrix}$. We highlight that Problem (38) and Problem (39) have equal optimal values and their optimal solutions differ from one to another by t, i.e., $\mathbf{a}^*=\mathbf{g}^*/t$. Next, we define the positive semidefinite matrix variable $\mathbf{G}\triangleq\mathbf{g}\mathbf{g}^H$ and drop the rank-one constraint on $\mathbf{G}$ by applying the semidefinite relaxation technique [20]. Then, the following relaxed problem can be obtained:

$$
\max_{\mathbf{G}}\frac{tr(\mathbf{P}_0\mathbf{G})}{tr(\mathbf{P}_1\mathbf{G})}
\tag{40}
$$
$$
\text{subject to } tr(\mathbf{P}_2\mathbf{G})\leq\gamma^2
$$
$$
tr(\mathbf{P}_3\mathbf{G})=1
$$
$$
\mathbf{G}\geq0
$$

It is worth mentioning that problem (40) is a quasi-convex problem due to the linear fractional structure of the objective function. Here, the Charnes-Cooper transformation technique [21] is exploited in order to transform Problem (40) into a convex SDP problem. We define the transformed variable $\mathbf{W} \triangleq \eta\mathbf{G}$, where $\eta \geq 0$ complies with $tr(\mathbf{P}_1\mathbf{W}) = 1$, and consider the following convex SDP problem:

$$\max_{\mathbf{W},\eta} tr(\mathbf{P}_0\mathbf{W})$$

$$\text{subject to } tr(\mathbf{P}_1\mathbf{W}) = 1$$
$$tr(\mathbf{P}_2\mathbf{W}) \leq \eta\gamma^2 \tag{41}$$
$$tr(\mathbf{P}_3\mathbf{W}) = \eta$$
$$\mathbf{W} \geq 0, \ \eta \geq 0$$

Based on Lemma 3.3 in [21], problem (40) is equivalent to (41) in the sense that they have equal optimal values and their optimal solutions differ from one to another by a constant η^*. In other words, once an optimal solution $(\mathbf{W}^*, \eta^*)$ to (41) is obtained, the optimal solution to Problem (40) can be easily obtained by $\mathbf{G}^* = \mathbf{W}^*/\eta^*$. Next, we focus on finding an optimal solution to Problem (39) with $\mathbf{G}^*$. First, we check the rank of $\mathbf{G}^*$. If $\mathbf{G}^* = \mathbf{g}^*\mathbf{g}^{*H}$ is rank-one, then $\mathbf{g}^*$ will be the optimal solution to Problem (39). If $rank(\mathbf{G}^*) \geq 2$, then there exists a vector $\mathbf{g}^*$ based on Lemma 1 such that:

$$\begin{cases} tr((\mathbf{P}_0 - v^*\mathbf{P}_1)\mathbf{g}^*\mathbf{g}^{*H}) = tr((\mathbf{P}_0 - v^*\mathbf{P}_1)\mathbf{G}^*) = 0 \\ tr(\mathbf{P}_i\mathbf{g}^*\mathbf{g}^{*H}) = tr(\mathbf{P}_i\mathbf{G}^*), \ i = 2,3 \end{cases} \tag{42}$$

Here, $v^* = tr(\mathbf{P}_0\mathbf{G}^{*H})/tr(\mathbf{P}_1\mathbf{G}^{*H})$ denotes the optimal value of Problem (40). As shown in Problem (42), $\mathbf{g}^*\mathbf{g}^{*H}$ is feasible to Problem (40) and the objective function evaluated at $\mathbf{g}^*$, i.e., $tr(\mathbf{P}_0\mathbf{g}^*\mathbf{g}^{*H})/tr(\mathbf{P}_1\mathbf{g}^*\mathbf{g}^{*H})$, is equal to the optimal value v^*; thus, $\mathbf{g}^*\mathbf{g}^{*H}$ is optimal for Problem (40) and $\mathbf{g}^*$ is the optimal solution to Problem (39). Then, the optimal solution to Problem (38) can be easily obtained by $\mathbf{a}^* = \mathbf{g}^*/t$. Algorithm 2 summarizes the procedure to find an optimal solution $\mathbf{a}^*$.

Algorithm 2. Procedure leading to an optimal solution to (38)

Input: $\hat{\mathbf{R}}_x$, $\mathbf{a}_U^*$ and γ
Output: An optimal solution $\mathbf{a}^*$ to (38)
1: solve the SDP problem (41), find an optimal solution $(\mathbf{W}^*, \eta^*)$ and the optimal value v^*.
2: let $\mathbf{G}^* = \mathbf{W}^*/\eta^*$
3: **if** $rank(\mathbf{G}^*) = 1$, **then**
4: perform an eigen-decomposition, $\mathbf{G}^* = \mathbf{g}^*\mathbf{g}^{*H}$
5: **else if** $rank(\mathbf{G}^*) = 2$, **then**
6: find $\mathbf{g}^* = D_2(\mathbf{G}^*, \mathbf{P}_0 - v^*\mathbf{P}_1, \mathbf{P}_2, \mathbf{P}_3)$
7: **else, then**
8: find $\mathbf{g}^* = D_1(\mathbf{G}^*, \mathbf{P}_0 - v^*\mathbf{P}_1, \mathbf{P}_2, \mathbf{P}_3)$
9: **end if**
10: **return** $\mathbf{a}^* = \mathbf{g}^*/t$

Finally, in order to avoid the ambiguity, the norm of the estimated actual steering vector $\mathbf{a}^*$ should be scaled back to $\sqrt{M}$:

$$\mathbf{a}^* := \frac{\sqrt{M}}{\|\mathbf{a}^*\|}\mathbf{a}^* \tag{43}$$

3.4. Proposed Robust Capon Beamformer

With the estimated actual steering vector $\mathbf{a}^*$, the weight of the proposed robust Capon beamformer can be formulated as:

$$\mathbf{w} = \frac{\hat{\mathbf{R}}_x^{-1}\mathbf{a}^*}{\mathbf{a}^{*H}\hat{\mathbf{R}}_x^{-1}\mathbf{a}^*} \tag{44}$$

The proposed robust Capon beamforming algorithm is summarized in Table 1. In the proposed method, the main computational complexity lies in the solution to the SDP Problem (27) and Problem (41), which requires a complexity cost of $O(M^{4.5}\log(1/\delta))$ with δ representing a prescribed accuracy [20], and both the specific rank-one decomposition and the matrix inversion operation require a computational complexity of $O(M^3)$. Consequently, the algorithm complexity of the proposed method is $O(\max(M^{4.5}\log(1/\delta), M^3))$.

Table 1. Proposed robust Capon beamforming algorithm.

1. Calculate the sample covariance matrix $\hat{\mathbf{R}}_x$ as (9)
2. Estimate the actual steering vector $\mathbf{a}^*$ in two steps:
Step 1: Estimate the oblique projection steering vector $\mathbf{a}^*_U$ via Algorithm 1
Step 2: Estimate the actual steering vector $\mathbf{a}^*$ via Algorithm 2 and normalize it as (43)
3. Calculate the weight $\mathbf{w}$ based on $\hat{\mathbf{R}}_x$ and $\mathbf{a}^*$ as (44)

4. Simulations

In the simulations, we assume a uniform linear sonar array (ULA) of 10 omnidirectional elements spaced half a wavelength apart receiving three Gaussian sources: the SOI from a direction $\theta_0 \in \Theta$ and two interferences from $\{\theta_1 = -40^\circ, \theta_2 = 50^\circ\}$. The interference-to-noise ratio (INR) is fixed at 40 dB. The presumed look direction is fixed at 0°. In addition to DOA mismatch, the array element calibration error and position error are also considered in all simulations. The element calibration error is caused by gain and phase perturbations, which are assumed to be uniformly distributed in [0.8,1.2] and $[-\pi/100, \pi/100]$, respectively. The element position error is assumed to be drawn uniformly from the interval [−0.05,0.05] measured in wavelength. Simulation results are averaged based on 1000 Monte-Carlo runs. The array element calibration error and position error change from run to run but keep fixed from snapshot to snapshot.

The proposed beamformer is compared with the SCB and robust approaches including the ESB [12], the Worst-Case beamformer [9], Zhuang's beamformer [16], Zhang's beamformer [17], and Wen's beamformer [18]. In Zhuang's, Zhang's, Wen's and the proposed beamformer, the ROI Θ is set to be $[-10^\circ, 10^\circ]$. Six principal eigenvectors of $\mathbf{C} = \int_\Theta \bar{a}(\theta)\bar{a}^H(\theta)d\theta$, whose corresponding eigenvalues contain 99.99% of the sum of all eigenvalues, are used to form the subspace $\mathbf{U}$. The values of the uncertainty level for the Worst-Case beamformer and Wen's beamformer are set to be $\varepsilon = 3$ and $\beta = 2$, respectively. Two cases with $\xi = 3$ and $\xi = 4$ are considered in the ESB and Zhuang's beamformer, where ξ is the estimated number of sources. The parameter $\gamma = 1.5$ is used for the proposed beamformer. CVX MATLAB toolbox [23] is used for solving the optimization problems in the Worst-Case beamformer and the proposed method.

4.1. Beampatterns

In the first simulation, we consider the resultant beampatterns of these beamformers. The actual DOA of the SOI is assumed to be $\theta_0 = 8^\circ$, that is to say, the DOA mismatch is 8° in this example. The number of snapshots $N = 500$ is taken. Figures 3 and 4 show the resultant beampatterns for SNR = 0 dB and SNR = −15 dB, respectively. The black dashed lines in Figures 3 and 4 represent the actual DOAs of the SOI and interferences. As shown, all these beamformers have deep nulls at the directions of interferences but only the ESB of $\xi = 3$ and the proposed method point their mainlobes towards the actual DOA of SOI in both SNR cases. The SCB, the Worst-Case beamformer and Zhuang's beamformer point their mainlobes towards the presumed look direction rather than the actual DOA of the SOI, and the SCB suffers from the problem of signal self-nulling when SNR = 0 dB, because the SOI is considered as an interference. The ESB of $\xi = 4$ also does not work well when SNR = 0 dB. Zhang's beamformer and Wen's beamformer fail to maintain the distortionless response towards the actual DOA of SOI when SNR = −15 dB.

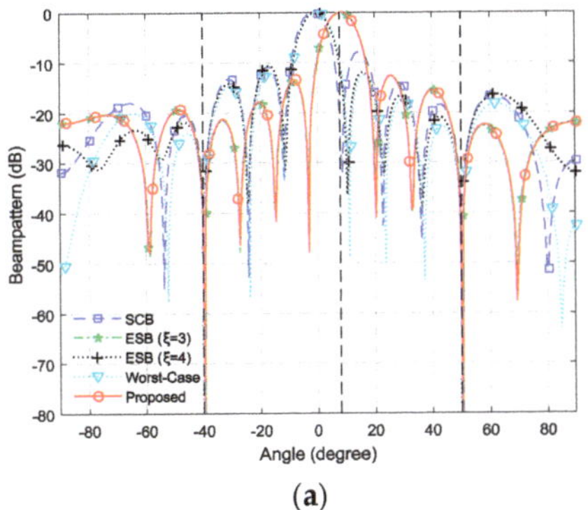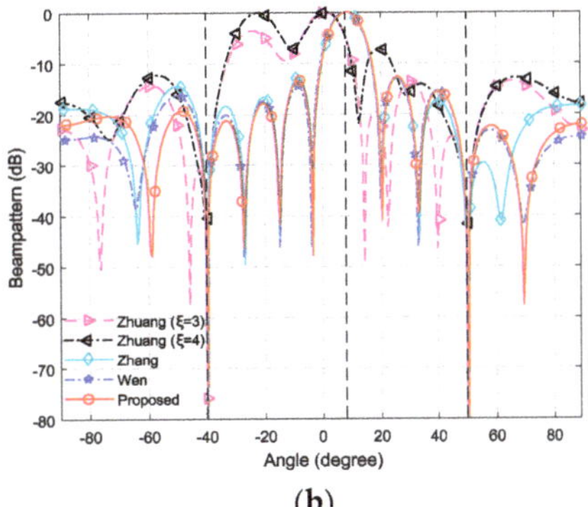

(a)   (b)

Figure 3. The resultant beampatterns for SNR = 0 dB. (**a**) The beampatterns of SCB, ESB, Worst-Case and the proposed beamformer; (**b**) The beampatterns of Zhuang's, Zhang's, Wen's and the proposed beamformer.

4.2. Output SINR versus Number of Snapshots

The second simulation aims to investigate the effect of the number of snapshots on the output SINR. We vary the number of snapshots from 20 to 1000 while the parameters of the SOI are kept unchanged. The output SINR versus the number of snapshots for SNR = 0 dB and SNR = −15 dB are shown in Figure 5a,b, respectively. In the figures, the black bold lines denote the optimal SINR. One can observe that the output SINR of the SCB, the Worst-Case beamformer and Zhuang's beamformer degrade significantly with a DOA mismatch of 8°. Zhang's beamformer and Wen's beamformer can provide sufficient robustness when SNR = 0 dB; however, their performances suffer from severe degradation when SNR = −15 dB. The ESB achieves high output SINR in both SNR cases, but it suffers from the overestimated number of sources. The proposed method achieves the best performance among all these beamformers.

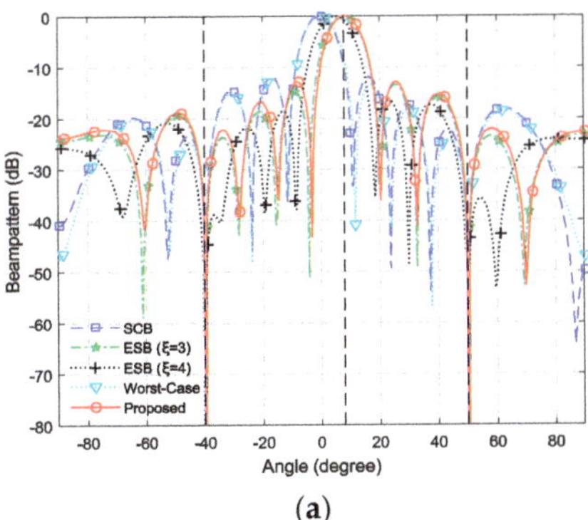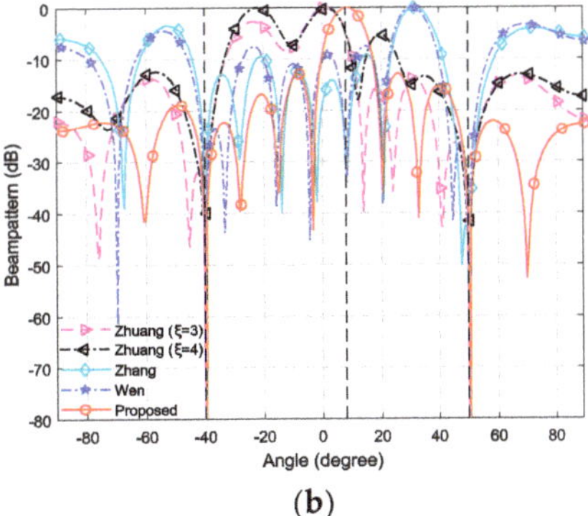

(a)   (b)

Figure 4. The resultant beampatterns for SNR = −15 dB. (**a**) The beampatterns of SCB, ESB, Worst-Case and the proposed beamformer; (**b**) The beampatterns of Zhuang's, Zhang's, Wen's and the proposed beamformer.

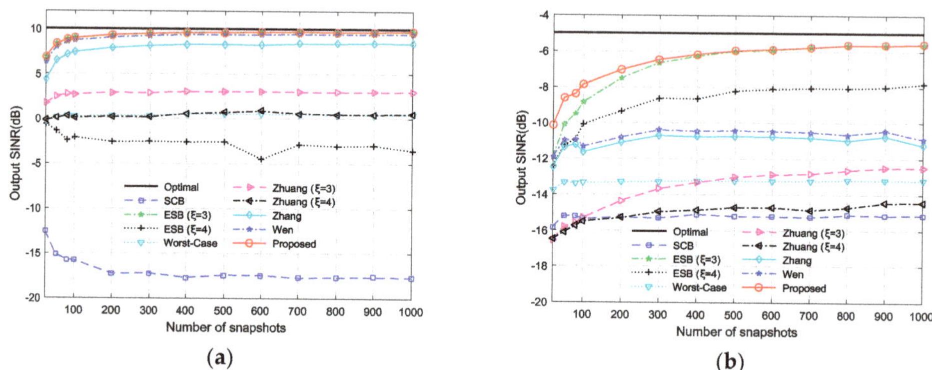

Figure 5. Output SINR versus the number of snapshots for: (**a**) SNR = 0 dB; and (**b**) SNR = −15 dB.

4.3. Output SINR versus Input SNR

Next, we vary the input SNR in order to investigate the effect of input SNR on the performance of these beamformers. Other parameters remain the same as the first example. Figure 6 compares the output SINR of the tested beamformers and the optimal SINR is denoted by the black bold line. The results indicate that the ESB of $\xi = 3$ and the proposed beamformer achieve the best performance among these beamformers. The SCB, the ESB of $\xi = 4$, the Worst-Case beamformer and Zhuang's beamformer do not work well with a large DOA mismatch. The performance of Zhang's beamformer degrades in the high SNR region. This is because it is sensitive to other array imperfections and the effects caused by the sensitivity will be enhanced when SNR is larger. Moreover, it also suffers from severe performance degradation in the low SNR region. Wen's beamformer has good performance in the high SNR region; however, its performance in the low SNR region is much worse than the proposed method.

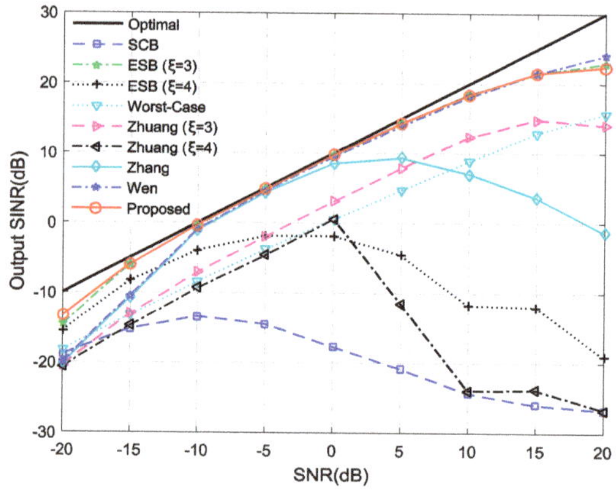

Figure 6. Output SINR versus input SNR.

4.4. Output SINR versus DOA Mismatch

In this simulation, we generally adopt the parameter setting in the first simulation and vary the DOA of SOI from −10° to 10°. That is, the DOA mismatch varies from −10° to 10°. The results for SNR = 0 dB are depicted in Figure 7a, and the results for SNR = −15 dB are depicted in Figure 7b. The corresponding optimal SINRs at SNR = 0 dB and SNR = −15 dB are represented by the black bold lines. It is clearly illustrated that the ESB of $\xi = 3$ and the proposed method outperform the other beamformers. While Zhang's beamformer and Wen's beamformer have sufficient capacity to provide robustness against large DOA mismatch when SNR = 0 dB, their performances degrade severely when SNR = −15 dB. The SCB is very sensitive to the DOA mismatch and it is more sensitive when the SNR is larger. The Worst-Case beamformer, the ESB of $\xi = 4$ and Zhuang's beamformer cannot provide robustness against a large DOA mismatch.

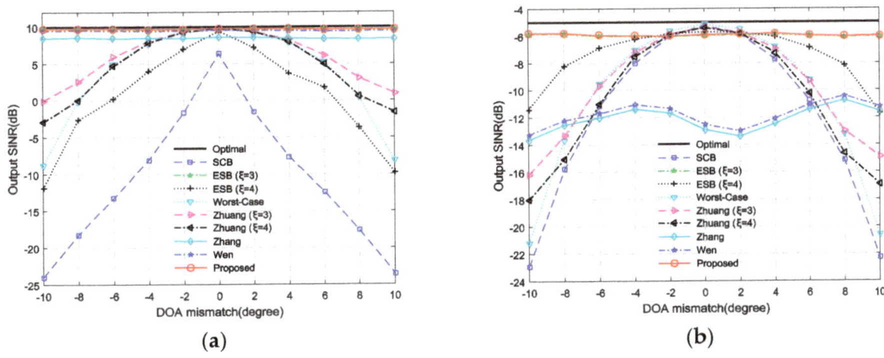

Figure 7. Output SINR versus DOA mismatch: (**a**) SNR = 0 dB; and (**b**) SNR = −15 dB.

4.5. Output SINR versus Parameter γ

Lastly, the effect of the parameter γ on the proposed robust Capon beamformer is considered. The SNR is fixed at 0 dB and other parameters are kept unchanged as the first simulation. Figure 8 shows the output SINR versus the parameter γ and the black bold line stands for the optimal SINR. One can observe that the performance of the proposed method remains steady and close to the optimal value in the range of $0.5 \leq \gamma \leq 3$. This implies that the performance of the proposed method is insensitive to the choice of the parameter γ, and thus we can set γ in a relaxed way.

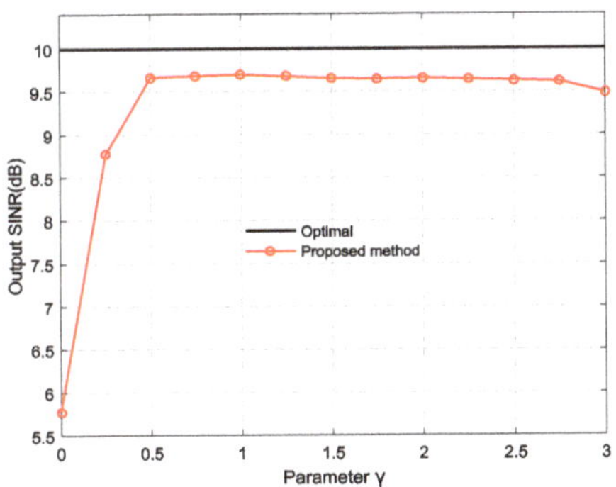

Figure 8. Output SINR versus parameter γ.

From the above simulation results, it is clear that ESB and the proposed method have superior performance on SINR improvement. The computational complexity of the ESB is $O(M^3)$, which is lower than that of the proposed beamformer. However, in comparison with the ESB which requires an accurate number of sources, the proposed beamformer only requires a relaxed parameter setting. Therefore, the proposed method is more suitable for practical applications of passive sonar systems.

5. Conclusions

A new robust Capon beamformer has been proposed in order to improve the robustness against the steering vector error dominated by large DOA mismatch for passive sonar. The proposed technique decomposes the actual steering vector into two components by projecting it obliquely onto a subspace. Based on this, the oblique projection steering vector is estimated within the subspace and then the actual steering vector is estimated within the neighborhood of the oblique projection steering vector. Nonconvex formulations for the two estimation problems have been solved by semidefinite relaxation, Charnes-Cooper transformation and rank-one decomposition theorem. The proposed beamformer can provide superior performance as compared with existing robust beamformers and only requires a relaxed parameter setting. Simulation results have been presented to demonstrate the effectiveness of the proposed beamformer in different hypothetical scenarios.

Author Contributions: Conceptualization, Software, Writing—Original Draft, Y.H.; Supervision, Writing—Review & Editing, N.Z.; Investigation, Funding Acquisition, G.L.

Funding: This work is supported by the National Key Research and Development Plan (2017TFC0306900), the Natural Science Foundation of China (11504064), the Heilongjiang Scientific Research Foundation for Returned Scholars (JJ2016LX0051), and the Open Found for the National Laboratory for Marine Science and Technology of Qingdao (QNLM20160RP0102).

Conflicts of Interest: The authors declare no conflict of interest.

References

1. Capon, J. High-resolution frequency-wavenumber spectrum analysis. *Proc. IEEE* **1969**, *57*, 1408–1418. [CrossRef]
2. Van Trees, H.L. *Optimum Array Processing: Part IV of Detection, Estimation, and Modulation Theory*; Wiley: New York, NY, USA, 2002; pp. 443–452.
3. Somasundaram, S.D.; Parsons, N.H. Evaluation of robust Capon beamforming for passive sonar. *IEEE J. Ocean. Eng.* **2011**, *36*, 686–695. [CrossRef]
4. Somasundaram, S.D.; Butt, N.R.; Jakobsson, A.; Hart, L. Low complexity uncertainty-set-based robust adaptive beamforming for passive sonar. *IEEE J. Ocean. Eng.* **2015**, *99*, 1–17. [CrossRef]
5. Somasundaram, S.D. Wideband robust Capon beamforming for passive sonar. *IEEE J. Ocean. Eng.* **2013**, *38*, 308–312. [CrossRef]
6. Li, J.; Stoica, P.; Wang, Z.S. On robust Capon beamforming and diagonal loading. *IEEE Trans. Signal Process.* **2003**, *51*, 1702–1715.
7. Li, J.; Stoica, P.; Wang, Z.S. Robust adaptive beamforming. In *Robust Adaptive Beamforming*; Li, J., Stoica, P., Eds.; Wiley: New York, NY, USA, 2005; Volume 3, pp. 91–93.
8. Li, Y.; Ma, H.; Yu, D.; Li, C. Iterative robust Capon beamforming. *Signal Process.* **2016**, *118*, 211–220. [CrossRef]
9. Vorobyov, S.A.; Gershman, A.B.; Luo, Z.Q. Robust adaptive beamforming using worst-case performance optimization: A solution to the signal mismatch problem. *IEEE Trans. Signal Process.* **2003**, *51*, 313–324. [CrossRef]
10. Li, J.; Stoica, P.; Wang, Z.S. Doubly constrained robust Capon beamformer. *IEEE Trans. Signal Process.* **2004**, *52*, 2407–2423. [CrossRef]
11. Beck, A.; Eldar, Y.C. Doubly constrained robust capon beamformer with ellipsoidal uncertainty sets. *IEEE Trans. Signal Process.* **2007**, *55*, 753–758. [CrossRef]
12. Chang, L.; Yeh, C.C. Performance of DMI and eigenspacebased beamformers. *IEEE Trans. Antennas Propag.* **1992**, *40*, 1336–1347. [CrossRef]
13. Hassanien, A.; Vorobyov, S.A.; Wong, K.M. Robust adaptive beamforming using sequential quadratic programming: An iterative solution to the mismatch problem. *IEEE Signal Process. Lett.* **2008**, *15*, 733–736. [CrossRef]
14. Khabbazibasmenj, A.; Vorobyov, S.A.; Hassanien, A. Robust adaptive beamforming based on steering vector estimation with as little as possible prior information. *IEEE Trans. Signal Process.* **2012**, *60*, 2974–2987. [CrossRef]

15. Fan, Z.; Liang, G. Self-steered robust adaptive beamforming. In Proceedings of the 2016 IEEE/OES China Ocean Acoustics (COA), Harbin, China, 9–11 January 2016.
16. Zhuang, J.; Manikas, A. Interference cancellation beamforming robust to pointing errors. *IET Signal Process.* **2013**, *7*, 120–127. [CrossRef]
17. Zhang, W.; Wang, J.; Wu, S. Robust Capon beamforming against large DOA mismatch. *Signal Process.* **2013**, *93*, 804–810. [CrossRef]
18. Wen, J.; Zhou, X.; Zhang, W.; Liao, B. Robust adaptive beamforming against significant angle mismatch. In Proceedings of the 2017 IEEE Radar Conference (RadarConf), Seattle, WA, USA, 8–12 May 2017.
19. Shahbazpanahi, S.; Gershman, A.B.; Luo, Z.Q.; Wong, K.M. Robust adaptive beamforming for general-rank signal models. *IEEE Trans. Signal Process.* **2003**, *51*, 2257–2269. [CrossRef]
20. Luo, Z.Q.; Wong, K.M.; So, A.M.C.; Ye, Y.; Zhang, S. Semidefinite relaxation of quadratic optimization problems. *IEEE Signal Process. Mag.* **2010**, *27*, 20–34. [CrossRef]
21. Mario, A.D.; Huang, Y.; Palomar, D.P.; Zhang, S.; Farina, A. Fractional QCQP with applications in ML steering direction estimation for radar detection. *IEEE Trans. Signal Process.* **2010**, *59*, 172–185.
22. Ai, W.; Huang, Y.; Zhang, S. New results on Hermitian matrix rank-one decomposition. *Math. Prog.* **2011**, *128*, 253–283. [CrossRef]
23. CVX: Matlab Software for Disciplined Convex Programming. Available online: http://www.standford.eud/boyd/cvx (accessed on 15 December 2018).

MDPI
St. Alban-Anlage 66
4052 Basel
Switzerland
Tel. +41 61 683 77 34
Fax +41 61 302 89 18
www.mdpi.com

Journal of Marine Science and Engineering Editorial Office
E-mail: jmse@mdpi.com
www.mdpi.com/journal/jmse